Thomas Bryant

Hunterian Lectures on Tension, as Met with in Surgical Practice

Inflammation of Bone and on Cranial and Intracranial Injuries

Thomas Bryant

Hunterian Lectures on Tension, as Met with in Surgical Practice
Inflammation of Bone and on Cranial and Intracranial Injuries

ISBN/EAN: 9783337162016

Printed in Europe, USA, Canada, Australia, Japan

Cover: Foto ©berggeist007 / pixelio.de

More available books at **www.hansebooks.com**

HUNTERIAN LECTURES

ON

TENSION, AS MET WITH IN SURGICAL PRACTICE.

INFLAMMATION OF BONE.

AND ON

CRANIAL AND INTRACRANIAL INJURIES.

DELIVERED BEFORE THE ROYAL COLLEGE OF
SURGEONS OF ENGLAND, JUNE, 1888.

BY

THOMAS BRYANT, F.R.C.S.

M.Ch. (Hon.) Roy. Univ. I.

VICE-PRESIDENT AND MEMBER OF THE COURT OF EXAMINERS OF THE ROYAL COLLEGE
OF SURGEONS; CONSULTING SURGEON TO GUY'S HOSPITAL; MEMBER
OF THE SURGICAL SOCIETY OF PARIS

LONDON
J. & A. CHURCHILL
11, NEW BURLINGTON STREET
1888

CONTENTS.

LECTURE I.

PAGE

On the Causes, Effects, and Treatment of Tension as met with
in Surgical Practice   1

LECTURE II.

On the Effects of Tension, as illustrated in Inflammation of
Bone and its Treatment   .   .   43

LECTURE III.

On Cranial and Intracranial Injuries   98

HUNTERIAN LECTURES.

LECTURE I.

ON THE CAUSES, EFFECTS, AND TREATMENT OF TENSION
AS MET WITH IN SURGICAL PRACTICE.

MR. PRESIDENT AND GENTLEMEN,—

When, through the kindness of my colleagues in the Council of this College, I was invited to accept the responsible position of Professor of Surgery and Pathology in this honoured Institution, I acceded to their request as a matter of duty, although with much diffidence, as I felt mistrustful of my power to bring before an audience, such as is wont to meet in this theatre, either material of sufficient importance to excite their interest, or to place it before them in a way sufficiently attractive to satisfy their critical requirements. To excite your interest, I have therefore selected a subject with which practical surgeons have long been familiar, and the importance of which they have recognised, but concerning which there is little or no literature—I allude to that of

1

Tension; and should I fail to make it sufficiently attractive, I have confidence that it will prove suggestive and tend towards some practical good. During the last few years the word "tension" has been freely used by both physicians and surgeons, although it has not been always employed with the same meaning. In my own student days it was rarely, if ever, heard; indeed, in a surgical point of view it had then but little significance. At the present time, however, we read and hear of it in many senses. The physician talks to us of arterial and muscular tension, and all admit that the word, as thus applied, carries with it deep meaning. The surgeon uses the term as applied to the pressure brought about by the distension or stretching of tissues by cystic or solid growths, by the extravasation of blood, and more particularly by what is far more common—the effusion of inflammatory fluids. I propose, therefore, in the following lecture, to invite your attention to these different causes of tension, and to trace their effects. I shall do this from the clinical point of view, under the conviction that some practical good may be derived from a full consideration of the subject, and with the hope that some light may thus be thrown upon the diagnosis and treatment of surgical disease.

With respect to the meaning of the word " tension " as employed in surgical work, and particularly in clinical work, it most frequently means the pressure brought about by the stretching or distension of tissue from either the growth of some neoplasm or the effusion of some fluid; tension, in this sense, meaning distension or the stretching of parts by a force acting from within—by centrifugal pressure, as it may be rightly termed. It is, however, applied in another way; that is, to the stretching of tissues which have been divided and brought together by sutures, the strain upon the sutures from the elasticity of tissues being the measure of the tension.

The effects of tension will be found to vary according to the nature of the tissue subjected to its influence. In one of an elastic kind, which yields readily under distension, the effects of tension are neither much felt nor well displayed, unless the expanding or distending force be carried to its full extent; whereas in a tissue which is unyielding and inelastic the mildest distending force is resented, and the effects of tension are forcibly demonstrated. Again, when the distending or stretching medium acts *rapidly*, the tension brought about in the tissues is severe, the symptoms associated with it are serious, and its effects destructive. On the other hand, when the distending, stretching, or straining medium acts *slowly*, tension is seen acting at a lower level, its symptoms are modified in intensity, and its effects qualified. As a general rule, the severity of the effects of tension, as well as the severity of the symptoms which characterise its different degrees, is found to turn upon the acuteness of its action and the elasticity of the tissues implicated. To this rule, however, there are exceptions.

The *ultimate* effects of tension upon any tissue turn, as already stated, upon the elasticity of the tissue and the rapidity with which the tension has been brought about; but they are invariably destructive. Its *immediate* effects, or mode of action, are primarily upon the circulation, particularly the venous; and the pressure from within of necessity tends to bring about—first a slowing of the capillary blood current through the stretched parts, and later on its stagnation, from which the death of tissue follows. When tension is very great, the venous and probably the arterial circulation through the tissues may be absolutely arrested. The nerves of the implicated tissues are at the same time stretched or pressed upon, and as a result pain is produced, and the severity of the pain is determined by the degree of pressure or stretching to which the nerves are subjected, and the character and quality of the nerve supply to the part. The pressure of tension, being centrifugal, acts all round. When the tension has been brought about by the effusion of inflammatory fluids, the effects described are aggravated, for the blood stasis which is well known always to exist in inflamed tissues is encouraged by tension; but of this later on. Where tension occurs in tissues which are not inflamed, inflammation is excited even when the tension is maintained at a low level. Where tension is more severe, destructive inflammatory changes rapidly supervene; but where it is most severe, the death of structure may result from tension without inflammation. Surgeons are familiar with many of these facts in their treatment of wounds; for it is to prevent tension and its evil effects that the drainage of deep, indeed of all, wounds is of such primary importance.

Symptoms of Tension.—The one subjective symptom to which tension gives rise is pain; and this is always found to vary with the degree of tension to which the tissues are exposed, and the quality and quantity of the nerve-supply to the part. Other things being equal, a low degree of tension is associated with slight pain, and a high degree of tension with intense pain. Pain under all circumstances has a common relation to tension. In parts badly supplied with nerves, or at any rate with sensitive nerves, there may be tension and yet no pain, even when from tension the vitality of the stretched tissues may be destroyed. An ovarian cyst, for example, may be so tense as to be deprived of life, and this change be unattended by pain.

The Diagnosis of Tension.—In superficial structures tension can, as a rule, be readily estimated by palpation. In deeper parts this may be difficult; in bone, or in the cranial cavity, it is impossible. An educated finger can, where palpation is applicable, be brought to diagnose degrees of tension, as indicated by elasticity and hardness; although where the tense tissues lie deep this may be difficult.

Where tension is not acting at its highest force, what is known as fluctuation may be made out. Where it is so acting fluctuation may not be found, only unyielding hardness of the parts implicated. The degree of tension of the part can, however, be determined by palpation, as in the eye, when the affected organ is compared with that of the sound side. The aspect of a tense tissue, moreover, helps diagnosis; its palpable enlargement as compared with the opposite and unaffected part, and its stretched appearance, being suggestive. In such a joint as the knee this condition can be well observed. But when the tense tissue is well covered with soft parts, as in a femur, the seat of periostitis, this observation cannot be made; but even there the enlargement of the part, and the engorged veins visible upon its cutaneous surface, are of diagnostic value as indicative of deep pressure. The diagnosis of tension of superficial, or comparatively superficial, structures can therefore be made with accuracy by observation and palpation; whereas, of deeper structures it can only be rationally inferred. Its existence can, nevertheless, be generally tested by surgical means, and its degrees measured. I must not, however, allow myself to dwell upon the diagnosis of tension under all circumstances, but with these general remarks, which have been made to clear the way, I will pass on to consider the effects of tension as brought about by the growth of tumours and the extravasation of blood, to those of tension, the result of inflammation, which probably forms the most important feature of the whole question.

On Tension from New Growths.—Where tissues are stretched or distended by a new growth of a non-infiltrating or innocent character, and its increase is slow, the effects of tension upon the parts which cover it depend greatly upon the character of the tissue in which it is placed and its capacity for yielding. Thus, if the tumour be a subcutaneous lipoma, there is rarely pain, as there is no tension; and where pain is present, it is probably due to the implication of some cutaneous nerve. Should the tumour be more deeply placed, as in the breast, which is covered by skin and fascia, and still be innocent, there is not pain of any significance so long as the tumour is small and does not give rise to stretching or distension of its coverings, and what pain exists will probably be of a dull kind. The skin over the tumour, however, as growth continues, changes in appearance, and its normal healthy aspect becomes congested. The engorged vessels are at first few, but later on many, so that at last the surface of the tumour assumes a congested leaden hue. The slow tension of the tissues thus brings about a gradual blood stasis, which passes into or induces the condition of inflammation, and subsequently of ulceration; tension, acting at a low level for a lengthened period of time, always bringing about, although slowly, destructive changes. Should a simple solid or cystic tumour originate in still deeper structures, and be covered with a dense fascia, a very different series of symptoms has to be described. As soon as the fascia becomes stretched or the deep parts distended, tension or centrifugal pressure is produced, and with it pain becomes a symptom of more or less gravity; the character and severity of the pain being determined by the sensibility of the parts involved, the degree of distension of the tissues, and the nature of the nerves involved or pressed upon. For example, a tumour situated between muscles loosely surrounded may, in its early stage, cause little or no pain, since the parts yield under distension; but as it grows and stretches its fascial coverings, pain appears and steadily increases. If growth still continues and the tension of the fascia becomes great, the pain from tension is not only severe, but it is aggravated by the backward or deep pressure of the tumour upon the nerve-trunks of the part in which it lies; the pain of local nerve pressure being added to that due to distension.

Again, should a tumour originate in the sheath of a nerve, pain is at once produced and rapidly increases in severity. The tumour, by its expansile growth not being able to bring about a yielding of the nerve-sheath, presses backwards upon the nerve-fibres, and causes suffering. To illustrate these points the following cases may be cited.

CASE 1. *Tumour in the temporal region beneath the temporal fascia and muscle; intense local pain from tension.*—Mrs. C—, aged thirty, came under my care in January, 1865, for a swelling in her right temporal region, which had been growing slowly for three years. It was hard and fixed to the temporal bone, and pressed forwards to the zygoma. The jaw was quite fixed, and could not be moved, apparently because of the stretched temporal muscle. The seat of swelling was the site of intense local pain, evidently due to stretching of the tissues. The lady subsequently died of exhaustion, after nine months of intense suffering.

CASE 2. *Tumour in a nerve-sheath (median); removal; cure.*—Miss P—, aged twenty-three, came under my care in August, 1876, with intense pain in the parts supplied by the median nerve of the right hand, contraction of the fingers and inability to extend them on account of pain, and a deeply-seated swelling, about the size of half a hazel-nut, above the anterior annular ligament in the median line of the forearm. The pain had been present for four or five years, but the swelling had only been noticed about six months. It had been mistaken for a ganglion and been punctured, but without relief. I regarded the case as one of neuroma in the median nerve, the tumour pressing upon the nerve exciting pain and forbidding extension. I therefore made a clean cut down upon the nerve, and turned out of its sheath a cystic growth the size of half a nut.* Immediate relief was given, and a good cure followed, the movements of the hand being perfect.

CASE 3.—In 1882 I removed from the upper cord of the left brachial plexus of a lady about fifty years of age a cystic tumour the size of a large hazel-nut, which had been growing for many months. It had caused intense pain down the arm, chiefly in the course of distribution of the musculo-spiral nerve, and from its hardness felt very like an exostosis. On exposure in the subclavian triangle, it was found to be in the upper nerve-cord of the brachial plexus, and within its sheath. On division of the sheath the cystic tumour rapidly enucleated, and this process clearly proved the tension to which the parts had been subjected, and explained the severity of the pain the patient had experienced. Dr. Goodhart reported that the growth seemed to be a blood-cyst. A good result followed the operation, and the lady is now well.†

When tumours originate in bone, there is, under certain conditions, severe pain, whilst under others there is but little. The presence or absence of pain and its degree are regulated by tension. For example, suppose a solid or cystic growth originates in the antrum of the upper jaw. As it grows it

---

* Prep. Guy's Museum, 1614⁵¹.
† Prep. 1614⁵ Guy's Museum.

tends towards the expansion of its bony casing, and gives rise to tension. As a result pain is induced. This, however, is not often very severe; for the thin bony wall, as a rule, gradually yields to the slow pressure from within, and in due time gives way, thus relieving tension. As a consequence, what pain may have been present, at once becomes ameliorated, and what is left is to be explained rather by the implication of the nerve-trunks of the part than by tension. When the shaft of a long bone becomes the seat of a cystic growth there is more pain, and this pain is from tension, since the wall of bone which surrounds the cyst is dense and unyielding to the centrifugally expanding pressure of the cystic growth. The pain in a case such as this is of the usual aching character, but it does not appear to be aggravated by the warmth of bed, as in inflammatory troubles. These points are well illustrated in the following case.

CASE 4. *Expansion of the shaft of the tibia from a cyst; trephining of the bone; recovery.*—Henry D—, a healthy man, aged twenty-four, came under my care at Guy's Hospital on January 6th, 1871, with considerable expansion of the centre of the shaft of the right tibia, which had been coming for two years. He had at times suffered much pain in it, but as a rule it had little more than ached after exertion. He had never had syphilis, and no history of an injury could be obtained. When seen, the bone was found to be expanded in its centre to twice the diameter of the normal bone. The surface of the swelling was much grooved by the cutaneous veins, and the integument over it was slightly œdematous. The parts were tender on firm, but not on gentle, pressure, and the seat of a dull, aching pain, which it is not stated was worse at night. On January 13th I cut down upon the swelling and trephined the bone with a large-sized instrument, and exposed, one inch from the surface of the bone, a cavity as large as a full-sized walnut, which was lined with membrane or granulation tissue, and contained serum more or less blood-stained, but no pus. It was not a hydatid. The bone cut through was denser than natural, and the periosteum over it was thickened. After the operation all pain ceased and a rapid recovery ensued. When I saw the man three years later he was quite well. The enlarged bone had contracted to its normal dimensions, and, beyond the scar of the operation, there were no indications of trouble.

On the other hand, should a solid sarcomatous growth originate in the centre of a dense bone in which there is but little cancellous tissue, such as the lower jaw, in which also there is a canal containing a sensitive nerve, tension in its highest force is the result. This was forcibly illustrated in a case of my colleague Mr. Cock, many years ago at Guy's.

CASE 5.—A man had a fibrous tumour growing in the dental canal of the lower jaw, and involving the inferior dental nerve. The agony this man suffered was excruciating, and the relief he experienced by the operation of trephining the bone and removal of the growth was almost magical. In this case the dense bone which surrounded the tumour resisted the effects of steady pressure, and was not absorbed as cancellous bone would have been. The effects of distension or tension upon the walls of the cavity containing the growth was consequently strongly marked and made manifest by pain, which was in its turn intensified by the compressing influence of the fibroma backwards upon the dental nerve. As an example of tension brought about by a slowly developing solid growth in unyielding sensitive structures, I am unable to adduce a better.

Tumours of bone do not, however, always distend bone and give rise to tension. Solid sarcomatous growths, which originate in the cancellous tissue of the long bones, rarely do so, at any rate to any extent; for the tumour, as it grows, by its constant pressure, steadily causes atrophy or absorption of the bone. And the enlargement of the cavity in the bone, going on *pari passu* with the increase of the tumour, neutralises the effects of tension, and reduces its influence to a low degree. The patient, consequently, under these circumstances, suffers but little, and what pain exists is rarely of a severe

or constant kind. As the tumour grows, however,
the little pain which may have been present disappears, for the bone after a time becomes perforated
by the growth, and tension therefore ceases to
exist. With its disappearance pain at once diminishes, and unless the growth presses upon nerves,
or structures well supplied with nerves, its progress is probably almost painless. It would appear,
therefore, that in the growth of tumours of the soft
parts, as well as of bone, the influence of tension is
one which cannot be ignored. As the chief cause
of pain, directly or indirectly, it is a potent factor,
and as a clinical symptom it is one of great diagnostic value.

With this brief reference to tension of tissues, as
induced by the growth of solid or cystic tumours,
or to what may be called slow tension, I must pass
on to consider tension as the result of sudden
effusion, and more particularly of blood, to which
class the term acute or rapid tension seems applicable. When a cavity or tissue becomes suddenly
distended from any cause, the amount of tension
which ensues is inversely proportionate to the yielding elasticity of the walls of the cavity or tissue
itself. The effects of the distension turn upon the
degree of tension that is produced; and the symptoms to which the tension gives rise vary with the
nerve supply to the part. Thus, in every-day practice, where blood is effused, as a consequence of an
injury, into subcutaneous tissue which is thin,

elastic, and capable of distension, as in the eyelids and scrotum, the patient experiences but little pain or other morbid sensation beyond a feeling of fulness, unless the hæmorrhage is great and the parts are stretched to their utmost. Whereas, should the effused blood be rapidly poured out into the skin of the ear, the junction of the nose with the upper lip, parotid region, or outer labium, parts which are not only sensitive but incapable of much stretching, the pain is severe, and this symptom is entirely due to the stretching or tension of the tissues into or beneath which the blood has been extravasated. Where no distension of parts exists, and as a consequence there is no tension, subcutaneous hæmorrhage may take place to a considerable extent and give rise to little local pain or other symptom than local swelling. The pain which attends subcutaneous hæmorrhage is in proportion to the degree of stretching or tension of the parts implicated. No tension means little pain; severe tension severe pain. When hæmorrhage takes place beneath the deep fascia which binds down tissues, pain is often severe; the distension of the fascia, the tension of its fibres, and the centrifugal pressure generally, being its cause. This is fairly illustrated in a case of simple fracture above the ankle, and in most sprains of joints attended with blood effusion. As the blood is absorbed, pain goes and repair progresses.

When we consider the subject of hæmorrhage into cavities, and the symptoms it produces, the same conclusions are arrived at. Thus, a patient may bleed to death from hæmorrhage into the peritoneal cavity, without manifesting any other symptoms than those which are described as general. There is no pain. The same may be said with respect to bleeding into the thorax. In neither of these cavities can tension, as a result of bleeding, be well produced, and as a consequence there is no pain. In intracranial hæmorrhage, where there is no yielding of external tissues, but only backward pressure, the effects of tension are very marked and destructive.

When bleeding, however, occurs into cavities which are within the range of distension, and in which, therefore, the effects of tension or centrifugal pressure may be felt or demonstrated, a different result has to be told. Thus, when a knee-joint is the seat of fracture of any of its bones, and as a consequence hæmorrhage takes place into the joint, the cavity may become so distended with blood, and thus tense, as to produce severe local pain and general disturbance, which, if not relieved, mny be followed by the destruction of the joint. The eyeball may be the seat of extravasation of blood into its anterior, middle, or deeper chamber, and its presence be unattended by any pain, unless the hæmorrhage be sufficiently extensive to distend the globe and give rise to tension. The tunica vaginalis of the scrotum may likewise become the seat of hæmatocele, and the amount of pain associated with it will turn upon the rapidity of the effusion and the mechanical distension or tension of the walls of the cavity into which the blood has been poured. The pain in all these cases, when it occurs, is clearly dependent upon tension.

When hæmorrhage takes place into deep, unyielding, and possibly sensitive structures, the degree of pain to which it gives rise can hardly be measured.

CASE 6.—Some twelve years ago I was asked by the late Dr. Remington, of Brixton, to see a fine young man, of about twenty-five, who, the day previously to my visit, had received a blow upon one of his testicles, and from the time of its receipt had suffered intense local pain, which opiates and local treatment had failed to relieve. When I went into his room the patient was walking about with his hands grasping his genital organs, and moaning with agony. In one testicle there was intense pain of a throbbing character, and the pain had since the accident been so severe that the patient said he had neither slept nor rested. I examined the scrotum, and failed to see any external signs of injury. The painful testicle was, however, larger than its fellow, and this enlargement was in the body of the gland. On careful manipulation, I made out a tense point in the body of the testis, and this point was painful. The man's temperature was normal, and there did

2

not appear to be any local heat about the testicle. I came to the conclusion, therefore, that in this case the patient was probably suffering from hæmorrhage into the body of the testicle itself, and that the intense pain which was experienced was due to tension. I consequently persuaded him to allow me to introduce a fine exploring trocar and cannula into the swelling; which I did, giving exit to a jet of blood or blood-stained serum which spurted out in every direction. With this spurt all pain ceased, and a rapid convalescence followed. That the pain in this case was due to the tension caused by the rapid effusion of blood into a sensitive gland encased in an unyielding fibrous covering—the tunica albuginea—there can be little doubt; and with the diagnosis the treatment was simple.

CASE 7. *Puncture of the testicle in the operation of tapping a vaginal hydrocele; hæmorrhage into the testicle producing pain from tension; operation and recovery.*—Mr. D——, aged twenty-one, consulted me in March, 1871, on the advice of Dr. Tyson, of Folkestone, for some affection of his left testicle, which had been swollen for ten years or more, but had only been painful since it had been surgically treated. One month before I saw him he went to a surgeon, who tapped the swelling and drew off a blood-stained fluid mixed with blood-clot. When the instrument was withdrawn the swelling steadily reappeared, and the testicle became the seat of intense pain, the pain passing up the cord. Since then the swelling had steadily increased, and the pain had been sickening. When I saw him the testicle was the size of a cocoanut, ovoid in shape, smooth, and very painful. It was not hot. Thinking the case might have been been a hæmatocele of the tunica vaginalis, although the extreme pain suggested something else, I at once explored the tumour by means of an incision, and found that it was a hæmatocele—but of the body of the testicle, and not of the tunica vaginalis. There had evidently originally been a hydrocele of the tunica vaginalis, for when I operated there was evidence of this fact ; but the testis itself had been converted into one large cyst containing broken-up and fœtid blood. I removed the whole organ, which was useless, and a good recovery ensued. In this case the testicle had doubtless been punctured when the hydrocele was tapped, and as a consequence hæmorrhage followed the puncture. The pain which came on after the operation of tapping, and which had not existed before, was clearly due to tension within the testicle brought about by the bleeding.

I have another case of tension brought about by the extravasation of blood, which I should like to relate since it is well in point.

CASE 8.—In 1877 I saw with Dr. Donald Hood, of Green Street, but then of Caterham, a gentleman

who in a hunting accident had ruptured his urethra,
and was suffering from retention and urinary extravasation. These symptoms were relieved by perineal
incisions and catheterism, and all went well for three
days, when bleeding from the urethra again appeared,
associated with agonising pain at the root of the
penis. On examining him through the wound,
which I then did, I found the bulbous portion of the
urethra very tense and exquisitely painful, and it
was clear that this tension of the part was due to
fluid. I accordingly made an incision into the bulb,
and with the incision what must be described as an
explosion at once occurred, for blood was scattered
in all directions. Copious bleeding followed, which
clearly came from the artery of the bulb. The
hæmorrhage was controlled and all eventually went
well. The tension of the bulb in this case from
arterial hæmorrhage was testified by the presence
of intense local pain, and it was proved by the
explosive sound caused by the division of the distended tissues, as well as by the scattering of blood
upon the operator and his assistants as soon as the
retaining capsule was divided.

Tension in these cases was acting at its highest
force, and, as the parts implicated were highly sensitive, severe pain was the result. Tension being
relieved in every case, pain was banished, and repair
went on in its normal way.

This accumulative evidence which I have brought

together, I think, therefore, points to one conclusion —that tension and pain as met with in cases of growths, expanding tissues; or of hæmorrhage into tissues and cavities, are closely related; in fact, that pain in these cases is the result of tension.

On Tension the result of Inflammation.—I will now pass on to consider the subject of tension in its relation to inflammation, and need hardly remind you that swelling is one of the classical symptoms of the process, and that this swelling is due to the exudation from the slowly flowing blood stream of corpuscular liquid which passes through the inflamed tissues, and which may lead up to blood stasis; this slowing of the blood current and consequential exudation forming Sanderson's central phenomenon of inflammation. Indeed, these elementary truths are such as every student and practitioner should ever have before him when considering any inflammatory affection, for they form the key to an intelligent appreciation of the local symptoms and pathological phenomena of every case. When effusion takes place, therefore, into any tissue which is inflamed, swelling follows, and the amount of swelling turns upon the character of the tissue into which it takes place. Where there is much connective tissue, and the parts are elastic and yield to distensile forces, the swelling is great; but as the resistance to it is not serious there is little or no tension; such an effect only follows extreme distension. Where the effusion is poured out into connective tissue contained in fibrous sheaths, or beneath or within fascial envelopes, as in joints, the distending force is great, and tension consequently is often found acting to a high degree; and where it exists in bone or in unyielding bony cavities, tension of the severest type is met with.

That local pain is a sign of tension all agree, although all will probably not consent to the opinion that tension is the cause of most pain. In inflammation, however, I think there can be little doubt as to the truth of this view, for pain and tension seem to be practically proportionate, the former being felt in exact proportion to the degree of the latter, and rising and falling with its force. Indeed, in a clinical point of view, I believe that the pain of tension should be regarded as a test symptom of inflammation. Where there is feeble tension in any inflamed part, the pain hardly exceeds that of uneasiness; where there is high tension, the suffering is severe; where tension is relieved by natural or surgical processes, pain goes. Under all circumstances, the presence of pain in any local inflammation may be accepted as a sign and indication of tension and a call for its relief. When pain is increased at night—that is, when the patient is warm in bed and the circulation is acting at its highest force—the probabilities are that the affection is inflammatory, and that such increase in the force of the circulation which is promoted by warmth tends towards the production of tension or its aggravation.

The *effects of tension* upon the inflamed tissues depend much upon its degree. When it is but slight, tension may tend only towards the maintenance of the inflammatory process; when it is more severe, it must help, not only to keep up, but also to aggravate, the action; and when it is acting at its highest force, there is no escape from the conclusion that it assists powerfully to bring about the destruction of the tissues which feel its immediate influence, and to encourage inflammatory action in such as are more remote. How it helps to bring about destruction of inflamed tissue is not difficult to understand, since it acts in tissues that are inflamed in the same way as it has been shown to act in such as are not so affected—that is, by helping to bring about blood stasis. In tissues that are inflamed the tendency to blood stasis invariably exists, so, with tension acting upon them, there can be no surprise that this blood stasis, with all its evils, is greatly encouraged. To say, as students are taught, that death of inflamed tissue from tension is due to the cutting off of blood supply to the inflamed part is a fundamental error. The blood is in the tissues, but it is stagnant and not circulating. The death of inflamed tissues from tension takes place, as it does in a strangulated part, from the stasis of its own venous blood. It is a form of static, not of anæmic, gangrene—of death of tissue brought about by the stagnation of the blood in the capillaries, and not by a want of blood supply.

I will now proceed to illustrate the effects of tension upon different tissues ; and first of all let me take as an example, a case of cutaneous erysipelas, and assume that it has attacked the skin of a part loosely connected with the deep tissues and capable of ready distension. The local sensation produced by the disease does not go beyond a sensation of stiffness—it does not amount to pain ; but let the inflammation spread to parts so placed as to be anatomically capable of but little, if any, distension, how different then are the symptoms complained of, and how soon inconvenience passes into pain, even of an acute kind. Let anyone who has had erysipelas, or any inflammation of the nose, upper lip, nape of neck, or ears, answer this question, and at the same time tell us the comfort he experienced when the tension was mechanically relieved. In this example of cutaneous inflammation the view that local pain is fairly due to tension is well supported, and the relief that is given to the symptoms by local treatment helps the conclusion.

Let me now draw an illustration of the effects of tension from inflammation of a deep structure bound down by inelastic tissues, such as is met with in a familiar example of thecal inflammation. What pain is at once experienced here from the first onset of the trouble, and how rapidly it intensifies ! What

destruction of tissue follows if relief be not given!
And what comfort and rapid recovery ensue if the
case be rightly treated! These facts are familiar to
us all, and yet I am rash enough to say, even in
your presence, that I am somewhat in doubt as to
whether the generality of surgical practitioners
realise their full meaning, and so conform their
treatment to the exigencies of the case as to give
their patient the best chance of a recovery. Let
me take this example of thecal inflammation, therefore, as a text, and enlarge upon it, first of all following its natural course. Inflammation attacks
the tissue; the capillaries of the inflamed part consequently become more or less thrombosed, and as
a result effusion takes place. The effusion is confined in a tight channel, the walls of which are but
feebly capable of yielding, and as a consequence the
fluid, pressing centrifugally everywhere, not only
violently stretches or distends the tissues in which
it is confined and gives rise to high tension, but also
seeks a vent by flowing in the line of the least
resistance along the course of the tendon up the
fore arm. From this pressure, if relief be not given,
death of tissues speedily follows, this result being
brought about, as already described, by the influence
of tension in encouraging the blood stasis which
forms part of the process of inflammation. With
the death of tissue suppuration within the thecal
channel takes place, rupture of the weakened thecal
structure follows, and in the end comes the external

discharge of the dead tissues, with probably many burrowing abscesses. A vast amount of destruction of tissue from tension has taken place quite unnecessarily, since by active and scientific early treatment a very different result could have been obtained. The cause of all this extensive trouble was practically due to tension, and if this had been relieved in its early origin recovery would have been possible, nay probable, without loss of structure.

The inference to be drawn from such a case as this, and from all others like it, is, that to relieve tension by giving vent to pent-up fluids in distended tissues is a primary surgical duty, and that this practice cannot be applied too early, may safely be dogmatically advised. At the present day amongst experienced surgeons it may be said that this practice is fairly universal; but is it? Is there not even now a lingering dread in some surgeons' minds of incising or even deeply puncturing an inflamed tissue before suppuration has commenced? And have not many of us heard, when an incision has been made into such inflamed tissues as those just mentioned, something like an observation of pleasure that a fluid like pus has been seen to flow, as if to justify the act? Whereas the surgeon's pleasure should be where his duty lies, to give vent to pent-up inflammatory fluids before suppuration or other destructive changes have taken place. In my own practice, I as a matter of routine cut down freely and immediately upon a finger the seat of thecal inflammation, and always congratulate myself that I have done so when only serum and blood escape, and no pus; for the existence of pus means destruction of tissue under most if not all circumstances, a result which should be avoided.

Again, let me give another example. A joint, let me say a knee- or hip-joint, becomes inflamed and acutely distended with inflammatory fluids. The synovial membrane as a consequence becomes tense from distension, and there arises the natural fear that disorganising changes in the joint will soon occur, if the tension thus produced be not relieved. The joint consequently is either aspirated or punctured with a tenotomy knife through a valvular subcutaneous wound, to relieve tension, and *some little* of the serous exudation is drawn off or allowed to escape into the connective tissue, where it will become absorbed. This little operation at once not only relieves pain by relieving tension, but it does more, for in a few days the fluid which was left in the joint disappears by natural processes, and convalescence is soon established.

An acutely inflamed bursa may be similarly dealt with, and an equally good result expected; and in the following case a like treatment of a hydrocele was followed by alike result. The case I give in the words of Dr. Bower Harrison, of Manchester.

Case 9.—" In March, 1878, I was myself the subject of hydrocele of about seven or eight months' duration. On the 20th instant, on getting out of a warm bath, I introduced an ordinary hypodermic syringe into the hydrocele, and, using it as an aspirator, removed about ten minims of fluid. A few drops of water remained on the spot where the syringe had been used. This dew continued to show itself for some time afterwards, whilst a perceptible softening of the whole tumour took place. In a few days the hydrocele had perceptibly diminished, and in a short time had wholly disappeared. It never recurred. In mentioning the case to my friend, Professor Lund, of Manchester, he expressed himself as much interested, saying that he felt convinced that the diminution of tension in the case of an imprisoned fluid greatly promoted its absorption. He further stated that he was in the habit of speaking of this to his class as an important *law*." I am quite convinced that our friend, Mr. Lund, is right in this matter. Indeed, I have for years taught this law myself. The withdrawal of some of the distending fluid from a tense synovial or serous sac not only relieves pain with the tension of the part, but at the same time frees the lymphatic, venous, and arterial circulation from the impeding effects of local pressure, and thus, by encouraging a more normal or healthy action of the vessels generally, tends towards the relief of the blood stasis, which is the main important pathological condition of inflammation, and thus helps towards recovery. In the same way the contents of an abscess may often become absorbed by natural processes, after a sufficient quantity of its contents has been withdrawn to relieve tension.

Again, glaucoma attacks the eyeball of a patient, and, whether acute or chronic, is attended by distension of the globe of the eye and extreme pain, the result of tension. Its treatment consists in an operation for the relief of tension, and without an operation in one of its forms recovery is considered by all authorities to be hopeless. " We must always bear in mind," writes my colleague Mr. Higgens, "that an operation to be successful must be performed early," the author meaning by " early " that it should be undertaken before the secondary changes in the structure of the eye, the result of distension, have been brought about. Surely this principle of practice, which is so binding and valuable in the treatment of an affection of the eye due entirely to tension, should be equally binding, as it would be equally valuable, in the treatment of all local inflammations in which tension has such a pervading influence.

Acute inflammation of the pulp of a tooth affords another illustration of the evil effects of pent-up inflammatory fluids upon surrounding parts, and of the severity of the local symptoms which must be attributed to such a cause. " In this case," writes Moon, " every factor for the production of agonising

pain is present; the distensible pulp, largely supplied with nerves, undergoes vascular engorgement
within an unyielding case—closed in at all parts
except at the aperture of exposure." The trouble,
if left unrelieved, soon ends in the death of the
pulp, and too often in the extension of the inflammation to the periosteum of the jaw-bone and surrounding parts, with its ultimate bad effects. How
many of these consequences may be avoided by
judicious treatment I must not now stop to inquire;
clearly the majority. Where the antrum of High-
more is involved in acute inflammatory trouble the
same conditions are present. Tension, and as a
result, pain, are prominent symptoms, and these are
only to be relieved by surgical action.

In the surgical affections of the ear many illustrations of the same truths may be found. In
furuncles and inflammation of the external meatus,
as is well known, severe pain is experienced, the
pain being the direct result of the unyielding condition of the skin and connective tissue in which the
inflammatory process originates, and the effects of
the tension being indicated by the throbbing pain,
the feeling of fulness in the part, and the general
febrile disturbance which in these cases is so often
present. In acute inflammation of the middle ear,
or in the acute grafted upon a chronic affection, the
evil effects of tension are forcibly illustrated, for in
such cases the walls of the inflamed cavity are chiefly
bony; and where this is not the case, the foramina,

which communicate, on the one hand, directly externally, and, on the other, indirectly internally, with the cranial cavity itself, are only covered with a membrane. When this cavity, therefore, is filled with inflammatory fluid, and the inflammatory process continues, the full effects of the centrifugal distending force upon the walls of the cavity, in bringing about tension on its yielding and un-yielding walls, is very marked. Indeed, it may be said without fear of contradiction that there are few local inflammatory affections which give rise to more severe local or general symptoms than otitis media. The severe and agonising local pain, radiating in all directions, the high fever and general constitutional disturbance, and too often the brain symptoms and complications which mark its natural progress, all point to the evil effects of pent-up inflammatory fluids, whilst the rapid influence of sound local treatment support a like conclusion.

Again, in inflammation and suppuration of the kidney brought about by the presence of a calculus or any obstructive cause in the urinary track, the same conclusions find full support. In these cases the lumbar or kidney pain—usually synonymous— turns almost entirely upon the tension or distension of the pelvis of the kidney. When, from any cause, the ureter of a kidney thus affected is obstructed, pain is caused or increased ; when the ureter opens and allows the free exit of pus, pain diminishes, or even disappears. The disappearance of the pain

depends entirely upon the unblocking of the ureter
and the consequent relief to tension of the pelvis of
the kidney.   On this account a patient with a dis-
organised kidney may go on for years, suffering
when obstruction to the flow of pus takes place, and
being comparatively easy between-whiles.

Indeed, in any suppurating cavity the same cause
and effect can be traced.   Retained pus always gives
rise to pain more or less severe, and its severity
turns upon the distension of the abscess cavity.
When the tension is brought about suddenly, and
is severe, the pain is great, as in acute abscess;
where it is produced by a chronic process, the pain
often is but little, tension being little.   The pain of
abscess generally turns almost entirely upon the tension of the walls of its cavity, and its severity upon the
rapidity with which the tension is brought about, the
anatomical surroundings of the part having its due
influence.   If further illustrations of the effects of
tension were needed, abundance might be adduced
from the surgery of inflamed periosteum and bones.
For my present purposes such are not required,
although in another lecture I propose to bring the
subject before you as a whole.

I must, however, as a final and possibly con-
vincing exemplification of the truth of all my pre-
ceding remarks, refer to *tension as seen in wounds*,
the result of either accident or operation.

In the treatment of wounds, and more particularly
of deep wounds, the evil effect of tension upon the

process of repair is well recognised; and it may, I think, be said that, if wounds are treated upon approved modern methods, the subject of the wound be healthy, his surroundings wholesome, and due provision be made against the possibility of any fluid exudation from the wound being retained, so as to give rise to tension, the surgeon may with confidence look forward to the reparative process being carried out without the slightest trace of inflammation, and without pain; whereas, if this important point of treatment be forgotten, disregarded, or inadequately provided for, the slightest trace of tension upon the margins of the wound, or of distension of its depths, will to a certainty show its evil effects, not only by preventing quick or any form of union, but also by changing the quiet physiological process of repair into the pathological which we call inflammatory, and thus set up a destructive in the place of a constructive process. This result ensues quite irrespective of the form of wound dressing that may have been employed.

I should like from this chair to repeat what I have been long teaching, and what modern surgery has done so much to inculcate, that repair and inflammation are not only not identical but incompatible; that repair is a physiological constructive restorative process, whilst inflammation is a pathological destructive one.

When inflammation attacks a wound that is healing by John Hunter's first intention, arrest of repair first appears, then disrepair, and the injured tissues will, when the inflammatory process has subsided, have to heal by granulation. Should, by some chance, inflammation attack a healthy granulating surface, the granulations will at once break down, and the molecular death of tissue, or the ulcerating process, take the place of what had been the reparative ; the constructive process in both cases having been destroyed by the inflammatory and exchanged for the destructive. In the repair of all parts which have been damaged by the inflammatory process, inflammation must cease before repair begins. I would that this view of repair and inflammation were more generally entertained ; it would help clinical surgery, for it would lead all surgeons, in their treatment of wounds, to avoid and guard against every outside influence that can possibly give rise to the over-action which we call inflammation, whether such be the introduction or germination of microbes from without, or the effects of tension from within.

I am not, however, here discussing the general causes of local inflammation. My object is simply to show that tension in wounds may either originate or keep it up ; where the tension is severe, acute inflammation will follow, as in tissues which are not wounded ; where it is acting at a lower level, it will still be destructive, although less so in degree. Should sutures of any kind be introduced to keep the edges of a wound together, and upon one or two the effects of tension be felt, at such points local inflammation will originate and pass on to ulceration. Whereas about the other sutures at which there is no tension, and yet which in other respects are similarly placed, no such action will be found to exist. And this result will occur quite irrespective of the nature of the sutures employed, and however carefully the antiseptic or other treatment of the wound be carried out. The local inflammation in these cases is clearly due to local causes, and to no other.

Again, if in a deep wound, blood or serum is allowed to collect in sufficient quantities to separate its adjacent surfaces, and by so doing give rise to tension of the surrounding tissues, inflammation will to a certainty follow, and such, will pass on to suppuration, if the tension be maintained. Whereas, if the tension be relieved soon after its appearance, the inflammation will soon subside, and the reparative process reassume its physiological work. The wounded surfaces, however, under both circumstances, are not likely to unite by quick or primary union, but by the slower process of granulation. In fact, the influence of tension upon the repair of wounds, both as a cause of inflammation as well as a source of its persistency, supports, if it does not prove, the following conclusions which the consideration of tension associated with other conditions has led me to draw.

1. That tension has a wide pervading influence in

clinical surgery, as well as a decidedly marked effect upon the progress of disease.

2. That it is the product of many causes, and that these, for clinical purposes, may be conveniently divided into the inflammatory and non-inflammatory.

3. That it stands foremost amongst the causes of pain, and in inflammatory affections it is probably the chief pain factor.

4. That where the causes are not inflammatory the tension to which they give rise will, if maintained for any time at a low level, or rapidly rising to a high level, excite inflammation in the tissues affected.

5. That, where the cause is inflammation, the tendency of tension is to keep up or intensify the inflammatory action, and strongly to encourage its destructive influences.

6. That tension in every degree has a destructive tendency, and the rapidity of the destructive process has a direct relation to the acuteness of the tension.

7. That, as in wounds the slightest degree of tension is injurious, so, in their treatment, the use of the drainage-tube, or due provision for complete drainage, is a point of such primary importance as to relegate to a secondary position the mode and character of the dressing which is employed, since a want of attention to the efficient drainage of a wound under every form of dressing is followed by the same result.

If these conclusions be true, and I am satisfied that in the main they are, two others ask for expression, the first being the value of local pain as a clinical sign of tension and an indication for local treatment; and the second the expediency, if not necessity, of relieving tension as speedily as possible under all circumstances. To both of these points I must briefly call your attention.

The Treatment of Tension.—If arguments were needed to support the view that the pain associated with a local inflammation is chiefly due to tension, they would be found in its treatment; for if the local inflammation be situated in an extremity—say a hand—the local pain will at once be relieved by such a simple measure as elevation of the limb; the elevation tending at once to empty the venous capillaries, and at the same time to diminish the force of the arterial blood supply to the inflamed parts, and thus to relieve the blood stasis and the tension resulting from it. On the same principle, the arrest of the flow of blood through the main artery of the limb by temporary pressure, or the more or less complete and permanent occlusion of its lumen by temporary or permanent ligatures, has been found to act beneficially. By such means pain is at once relieved, and other advantages at the same time secured. Local venesection, leeching or cupping in either of its forms, must also act in the same way, by relieving tension; the local abstraction of blood from the seat of an inflammation or its neighbourhood so emptying the congested vessels in the outer zone of an inflammatory centre as to encourage the circulation of the part generally, and by so doing give relief to the overgorged and dilated capillaries of the centrally inflamed tissues. The local loss of blood, even when very limited in extent, relieves tension, and with it its diagnostic symptom, pain. In the same way, it is probable that cupping, dry or moist, and counter-irritation, have a beneficial tendency, and that these means, by drawing blood to a part not too remote from the seat of inflammation, help to relieve the congested capillaries in the focus of disease, and thus to favour the circulation through them ; this favouring of the circulation tending, of necessity, to discourage blood stasis and hyperæmia, to relieve pain, and to help the natural powers to bring about a cure of local inflammation by resolution.

In the inflammation of soft parts, where the stretched tissues are elastic, their natural resiliency may have a beneficial influence towards the re-establishment of the circulation and the favouring of a recovery by resolution; whereas, in the inflammation of the more inelastic and unyielding structures, such as the eyeball, middle ear, joint capsules, periosteum, and bone, no such help can be counted upon ; for when these parts become the subject of acute or chronic inflammation, and consequently the seat of tension, such tension and inflammation, unless relieved by surgical art, will continue, and thereby bring about permanent or destructive changes in the inflamed parts, the result being in one case thickening and consolidation of structure, and in the other suppuration, ulceration, or death of tissue. It would therefore appear that in the treatment of local inflammation the relief of tension is an all-important point of practice to be followed, and that the means which in any individual case can best fulfil this object will be the right ones to adopt. In some cases the relief may be brought about or helped by position, in others by pressure. In acute cases it is only to be efficiently produced by a puncture, incision, drill, or trephine into the tense tissues or cavities; whereas, when the soft and elastic structures of the body are the seat of trouble, local venesection, cupping, or the application of leeches may suffice, and it would be well if such means were more frequently employed. But wherever tension may be found, and in whatever tissue it may be situated, means must be employed for its relief; since, so long as it exists, the causes which led to it will probably continue, unless by the natural progress of the case tension becomes relieved by the destruction of the structures which are under its influence, or natural processes are sufficient to bring about a cure.

Where pain is present as an indication of the existence of inflammatory tension, the surgeon has a guide of great value to help him to the right treatment of his case, and it is one which he should rightly use; although where pain is absent or but feebly expressed, there is the same necessity to relieve tension where it exists, for tension may be present even with distension without giving rise to pain ; as, for example, in certain forms of peritoneal inflammation. In these cases, however, the same line of treatment is called for, and the same principle of practice is applicable as in others to which attention has been drawn ; for, with an abdomen full of fluid, which is pressing in all directions upon veins, lymphatics, arteries, and viscera, with an embarrassed circulation, and an arrest of all physiological action of the organs implicated, how can any curative action be expected to take place ? Let the distended abdominal cavity be relieved by the evacuation of its contents ; let some or all of the inflammatory products which may retard repair be washed away and the circulation be disembarrassed ; let the viscera be freed from all pressure, and the physiological action of all these important parts be allowed to have fair play, and then recovery may be brought within the bounds of possibility if not confidently anticipated. The good results which have followed the modern treatment of peritonitis by laparotomy and irrigation tend to support this view, and it is, I believe, by carrying out the principle I have feebly indicated that such successes are to be explained. In support of this view I would allude again to the partial withdrawal of fluid from tense cavities for the relief of tension, and the result which commonly follows the practice, for it cannot but be acknowledged that this action is often speedily followed by the absorption of the liquid which is left, and recovery from the local trouble. The partial withdrawal of pent-up fluids apparently so frees the venous, lymphatic, and arterial circulation of parts which have by the pressure been previously embarrassed, as to be followed by their full physiological action and the progress of repair towards a cure. In many a large hæmatoma I have despaired of its absorption and a natural cure till I had withdrawn some of its serous contents and removed tension, when a rapid recovery followed.

In not a few inflamed knee-, hip-, and other joints made tense by inflammatory effusions, and so tense as to excite a fear of loss of structure, has a rapid recovery followed a careful aspiration sufficient to relieve tension, or a subcutaneous puncture of the joint. Even in large abscesses, and particularly chronic abscesses, the same result may be chronicled, that is, the rapid absorption of what may have been left in the abscess cavity after the withdrawal of only a part. And the explanation of these results can probably be found in what has been already said.

I have thus, Mr. President, laid before you—although, I fear, imperfectly—many, if not most, of the clinical data connected with what has been called tension, or centrifugal pressure. I have pointed out its causes and its effects, and more particularly the great value of pain as a significant sign of most forms of tension, and especially of inflammatory tension. I have attempted also to bring out from these facts a certain principle of practice, the value of which I do not think I have exaggerated. If I were to test the value of the practical principle I have brought before you by my own personal experience, I should place it very high, for I can assure you that I have for years acted upon it and not found it wanting. To carry it out fully, we must, however, return to former lines of treatment, and employ more frequently than we do the practice of leeching in local and comparatively superficial inflammations, and of puncturing, incising, drilling, and trephining in bony and subfascial inflammations ; and these means should be employed early in every case. Should we employ counter-irritation, we shall do so on a kind of principle which is intelligible, and we shall therefore be more likely to employ our measures with success. To illustrate, however, what I have said to-day more fully and completely, I must go to the diseases of the bones and periosteum ; and it is to these that I hope to draw your attention in my next lecture.

LECTURE II.

ON THE EFFECTS OF TENSION, AS ILLUSTRATED IN IN-FLAMMATION OF BONE AND ITS TREATMENT.

MR. PRESIDENT AND GENTLEMEN,—

My last lecture was devoted to the subject of tension generally, and I trust I was able, with sufficient clearness, to satisfy you that tension of tissues has a widely spreading influence on the progress of disease, particularly of inflammatory disease. I pointed out what seemed to me very clear, that tension is a common factor of pain, and that in all inflammatory affections the pain of tension is not only a significant symptom of its presence, but a fair measure of its intensity. I showed how tension may and frequently does originate inflammatory action, more particularly in wounds, and that this effect takes place under every form of wound dressing. I demonstrated how in the soft tissues of the body, acute tension, or tension acting at a high level, and allowed to take its course, always ends in the destruction of the tissues implicated; and pointed out that where tension was acting for any time at a low level in either hard or soft tissues, permanent change of structure, tending towards their destruction, may be looked for; and I concluded that the most satisfactory and certain method of arresting acute as well as chronic inflammatory processes is by relieving tension.

In my present lecture I hope to enforce all these conclusions by considering the effects of tension in bone when in a state of inflammation, and I do so in the belief that it is in the inflammation of bone and its fibrous covering that the effects as well as the treatment of tension are best illustrated. The key to this conclusion is probably to be found in the vascular supply to the bone, and it will therefore be well to devote a few moments to its consideration. I have no intention, however, to weary you with any long anatomical description of bone and its covering. For our present purpose it is sufficient that we recognise the close vascular connection which exists between the compact tissue of bone and its periosteal envelope, as well as the free anastomosis of the vessels which supply the medullary canal of bone with those of its compact substance. It is likewise important to keep in view that this vascular connection applies as much to its venous as to its arterial supply. The arrangement of the capillaries of the bone also must not be forgotten, for these pass through fine bony canals, which are incapable of dilatation or contraction, and flow into veins similarly arranged. The veins, moreover, have pouch-like dilatations or sinuses in their course, are deficient in muscular fibres and valves, and in all such internal arrangements and external surroundings as in the softer structures help towards the free circulation of the blood. In fact, the circulatory system in bones is so arranged as to favour blood stasis. With such anatomical arrangements it becomes, therefore, clearly intelligible why an inflammatory action which may have originated in one part of the bone—let us say in the endosteum—so soon involves the outside compact bone structures; why inflammation of the periosteum which envelops the bone is so frequently complicated with inflammation of the bone itself; and, last but not least, why when bone is inflamed and there is the necessary slowing of the blood current through the capillaries and veins of the inflamed parts, it is as rare as it must be difficult for natural powers, unaided by art, to bring about a cure. Thus it is, therefore, that we find, when acute inflammation attacks the periosteum, that death of the bone from extension of the inflammatory process is a common consequence; that acute inflammation of the medullary cavity of a bone (osteo-myelitis, endostitis, ostitis), as a rule, spreads to its compact walls, and ends in more or less destruction of the bone; and that, where this result does not follow, the formation of abscess or abscesses in the affected bone, with partial necrosis, is the termination to be expected. For, when an acute inflammation has attacked a bone, whether as an extension from the periosteum or otherwise, its termination may be either in necrosis, partial or complete, or in the formation of an acute abscess or abscesses; the abscess cavity containing not only pus with the molecularly destroyed bone tissue, but as often as not a distinct sequestrum; for molar and molecular death of tissue with suppuration are the common results of acute inflammation of bone as of softer structures.

If we pass on to consider the effects of *chronic inflammation of bone*, the same want of recovery by resolution has to be recorded, with the same tendency for the inflammatory action to bring about destructive changes. An inflammatory process, once started, goes on slowly but surely, although possibly with intervals of relief and exacerbations; but it ends either in some abscess in the bone, the death of a small portion of bone with the condensation of more, or the sclerosing of the whole bone. Examples of all these terminations are both numerous and good in our own and other museums. It may be added, however, that where sclerosis of bone is met with, there is usually some degree of destruction of bone associated with it, either in the shape of limited necrosis or abscess; and I have reason to believe that if all apparently merely sclerosed bones were carefully divided into sections, either a small necrosed sequestrum or chronic abscess in the bone would be very generally discovered, the sclerosing process being due to persistent irritation kept up by the abscess or sequestrum. I have frequently myself found this, and consequently acted upon the

knowledge, and I may add with advantage.   In a case
to be recorded later on, this fact is well illustrated,
and in another case under Mr. Cock, which took place
in 1856, in which the limb of a man, aged thirty-
seven, was amputated high up in the thigh for enor-
mous enlargement of the femur of many years'
standing, and in which the bone was like ivory, a very
small piece of necrosed bone was found in its central
canal, unattended by any external suppuration.

To find a full explanation of this fact—that inflamed bone and its periosteal covering, as a rule,
ends in some destructive change—we must go to
tension, for by universal consent surgeons acknow-
ledge that the death of an acutely inflamed bone is
brought about by such a cause, and also, that the
intense pain which is connected with it, is, without
doubt, due to the inability of the bone tissue to
yield in any degree to the distending influence of
inflammatory hyperæmia and its consequences.   And
I trust I may find many to share with me the belief
that it is this same tension in subacute and chronic
inflammation of bone which not only keeps up the
process and prevents recovery by resolution from
taking place, but also tends to bring about the destructive and sclerosing changes in bone with which
we are all familiar.   Why it is that, in bone, tension
in any of its degrees is so fruitful of harm, must be
explained by the anatomical facts relating to its
circulation to which I have drawn your attention,
and the difficulty under which unaided natural pro-

cesses must consequently labour to enable the circulation of the bone to overcome the tendency to capillary blood stasis which forms the essential pathological phenomenon of inflammation, and upon which the life or death of the bone depends. For in bone, as there are neither in the normal arrangement of its capillaries and veins, nor in its surroundings, any such provisions as are found in soft tissues to favour the flow of blood in its usual physiological condition, so under the influence of disease, and more particularly of inflammation, in which the slowing of the blood current and capillary blood stasis are the invariable consequences, the difficulty of restoring the circulation to its normal condition must of necessity be infinitely greater. Under these circumstances the practical inference to be drawn is this—that it is the surgeon's duty, without reserve, when inflammation of bone and its covering can be made out to exist, to do what lies in his art to relieve tension; and it will be my object in this lecture to enforce this view.

As a preliminary point of great importance, however, I would draw your attention to the value of pain as an indication of the presence of inflammation of bone, such pain appearing to me to be the result of tension under all circumstances. In acute disease this is admitted by universal consent, and in the more chronic forms it has a no less general influence. The pain in its nature may not always be severe, or in the bulk of cases capable of being described as more than an ache, but this ache will probably at times be intensified by walking if the trouble be in the lower extremity, or by hanging the arm down during the day if in the upper limb —by anything, in fact, that helps blood stasis in the part. It is also, always, worse at night, when the circulation of the blood through the part with that of the body generally is encouraged by the warmth of bed; indeed, at these times the pain may be severe—very severe. Such pains in bone, increased at night, have been commonly believed to be associated with syphilis, and have been called osteo-copic pains. Such a conclusion may, however, be wrong. The pains, nevertheless, are always bone pains, and as a rule due to inflammation. Occasionally, indeed generally, there is increased local heat in the bone affected, although in deeply placed bones there may be a difficulty in detecting it. In such a bone as the tibia the increased heat which is present is very marked if rightly tested; and what I mean by the term "rightly" is as follows. The limb should be exposed with its fellow for purposes of comparison, and the palm of the surgeon's hand placed—if the suspected inflamed bone be the tibia—above the knee-joint, and allowed to rest there for a few seconds to appreciate the normal temperature. The hand should then be passed *slowly* downwards over the suspected bone, say, to the ankle; and during this action any increase in local temperature, even

4

if localised to the space of half-a-crown or less, may
be at once detected. The sound limb should then
be examined in the same way, and the sensations
acquired by the two hands compared. The test takes
less time to make than it has taken me to describe. It
is far better than simply grasping the suspected
bone with one hand and the sound one with the
other, the method in which the examination is
usually conducted. By the same method, even if
the femur be the bone affected, the increase in temperature can be detected. I trust I may be par-
doned for this digression. I have, however, no
desire to enter into the diagnosis of these cases
generally, and have dwelt upon the diagnostic value
of pain more to draw your attention to its use as
indicating the presence of inflammatory tension in
bone than for any other purpose, for my object is to
show you that this tension must be relieved, if any
help towards cure can be entertained. To a
certain extent most surgeons recognise the truth of
these remarks, and more particularly in their treatment of acute periosteal inflammation. For they
not only act upon the principle I am anxious to
enforce, but also teach their pupils to make an incision down to bone as soon as the diagnosis of acute
periostitis is sufficiently definite. This incision is
called for primarily with a view of relieving subperiosteal tension and the hyperæmia of the probably
involved bone; and secondarily, in order to guard
against the dangers, which should always be recog-

nised, of the inflammatory effusions burrowing beneath the periosteal coverings of the epiphyses, or through the epiphyses themselves into the neighbouring joints, and so originating a suppurative synovitis. No words that I can use would be too strong to support this practice, for it is without doubt the only reasonable one.

When acute ostitis, endostitis, or osteo-myelitis attacks the shaft of a long bone, and still more when it originates in the juxta-epiphysial cartilage or in the epiphysis itself, the necessity of making a free external incision is likewise a fairly well recognised practice. It is here adopted for the same reasons as have rendered it necessary in acute periosteal inflammations; but in these cases it should be regarded as a means only to a more important end—that is, to give room for the surgeon to relieve the tension in the bone itself, either by drilling, trephining, or otherwise freely exposing the centre of the bone in which the inflammation originated, or to which it may have spread. This operation is doubtless the only one which acts favorably upon the disease, in either arresting its course or limiting its destructive tendencies. But I would wish to go a step farther and urge that active local means to relieve tension should, as a principle of practice, always be immediately taken when such has been brought about by acute inflammatory processes, whether the periosteum or endosteum be its primary seat; and that the surgeon need not,

before he acts upon this principle, wait, and in so
doing lose valuable time, whilst a definite or exact
diagnosis is being made.　Indeed, in a clinical point
of view, acute periostitis and endostitis had better
be regarded as identical affections, since in both
there is, pathologically, rapid effusion and extreme
tension of tissue, which very quickly bring about its
death, and, clinically, deep swelling, severe local
pain, and marked constitutional disturbance.　The
general symptoms are frequently so severe as to
raise the suspicion of the patient being the subject
of an acute general rheumatic rather than of an
acute local affection, as proved by the fact that in
hospital practice many of these cases are primarily
admitted into the physicians' rather than the surgeons' wards.　Should joint symptoms coexist with
acute epiphysial inflammation, or follow diaphysial
ostitis as a result of burrowing of the inflammatory
fluids beneath the periosteum into the joint, the
surgeon should act more energetically, and he need
not despair of seeing the joint symptoms disappear
if relief can be afforded to the inflammation in the
bone and a vent given to inflammatory products.　If
the joint has not suppurated, a perfect recovery may
be anticipated ; but if this consequence has ensued,
a recovery with anchylosis should be expected.
The surgeon should remember that the destructive
effects of tension in acute periostitis or endostitis is
in most cases worked out within a week from the
onset of the inflammation, so that, if good is to be

gained by treatment, it is all-important that such should be applied early. He should likewise recognise the fact that by drilling or trephining inflamed bone necrosis is not produced; the operation may fail in preventing such a result taking place, but it will invariably tend towards its prevention and limitation. The incision down to bone in acute inflammation of the periosteum cannot be made too early; and in acute ostitis the same incision made as a preliminary measure to enable the surgeon to drill, trephine, or lay open the bone cavity itself, should evidence of its inflammation be present, can only tend towards good. If the periosteal incision be made in the early stage, in which blood-stained serum or blood alone is effused beneath the periosteum, the surgeon may entertain a reasonable hope that recovery will take place without necrosis; but if pus be present the chances are that more or less death of bone will ensue. In exceptional cases, however, a better result may be obtained. The following case of hæmorrhagic acute periostitis very forcibly illustrates all these points, with others, and is consequently worthy of attention.

CASE 10.—A boy aged nine came under my care in November, 1876, with symptoms of acute inflammation of the bones of the left leg and ankle, as indicated by great swelling, redness, and tension of the soft tissues over the bones, and intense pain in the part. His temperature was 104°, pulse 140.

These symptoms were said to have followed a kick
received three days previously. As soon as I saw
him a free incision was made over the fibula, where
the tension seemed to be the greatest, and the
periosteum was opened with a distinct explosive
sound, the fluid from beneath being scattered in all
directions. This fluid was, however, pure blood,
clots being subsequently syringed or washed away.
The bone was completely stripped of its periosteal
covering. Two months later a large piece of necrosed
bone exfoliated and was removed from the fibula,
and the limb recovered. On the fourth day after
the boy's admission into the hospital with acute
periostitis of the left fibula, which had been attributed to a kick, the right leg became affected in
precisely the same way; the temperature rose to
102·6°; the pulse to 144. The affection was treated
by a free incision over the fibula, as in the other
limb, but within twenty-four hours of the onset of
the symptoms, and blood escaped with the same
explosive sound as on the former occasion from
beneath the periosteum, and the bone was left freely
exposed. No necrosis, however, ensued. Five
days after his admission it was evident that the
periosteum of the tibia of the right leg was involved
with the fibula, and an incision was made directly
down to the bone, from which pus freely escaped.
The local and constitutional symptoms at once subsided, but necrosis of the tibia subsequently followed, and a sequestrum was removed at the end of

two months. A good recovery afterwards took place. In this case it is worthy of note that in the acute inflammation of the periosteum of the fibula, which came on whilst the boy was in bed, and was treated within twenty-four hours of the onset of symptoms by a deep incision, no necrosis followed. In the opposite limb, where the same trouble attacked the same bone and the incision was not made till the third day, necrosis ensued; and in the tibia, where suppuration was allowed to take place before the tension was relieved, the same result had to be recorded. The fluid that was effused beneath the periosteum of both fibulæ was blood, and in both limbs the tension of the periosteum was so great that its incision was attended with an explosive sound. No better case could be brought before you to illustrate the value of an early and free relief of local tension than the one I have briefly read, and certainly no better example could illustrate the benefit of action and early treatment in checking and limiting disease.

In the following two cases the periosteal incision was not made until suppuration had occurred; but in neither of them did necrosis follow. Such a possibility is consequently a strong inducement for the surgeon to interfere, even in cases which are unpromising.

CASE 11.—A boy aged ten (David K—) came

under my care in July, 1877, with his left thigh and knee-joint enormously swollen, tense, and tender. He was also in a state of high fever, with a temperature of 101·2°. The local trouble had come on two days previously, with pain and swelling, and this had followed a fall upon the knee two days earlier. The diagnosis made was acute femoral periostitis, with effusion into the knee-joint from contiguity. An incision was therefore made into the outer side of the thigh, between the tendon of the biceps and ilio-tibial fascial band, down to the femur, when six or more ounces of pus escaped, the femur being found stripped of its periosteal covering for some inches. The abscess cavity was washed out and drained. Great and immediate relief followed this operation, and a steady convalescence in about six weeks. There was no dead bone as a result of the inflammation, although pus had formed in two days from the origin of the trouble.

CASE 12.—In this case a like favorable result took place. A boy aged four (W. P—) came under my care in 1876, with a deep-seated, tender, and painful swelling involving the centre of the left femur, which had speedily followed a fall downstairs nineteen days previously. Pain and swelling had appeared a day or so after the fall; but for a week the child limped about, although he was ill. When I saw him a deep periosteal abscess evidently existed, which I laid open, evacuating pus and exposing bone. Rapid recovery ensued, and without loss of tissue.

If I am not wearying you, I should like to quote the leading facts of other cases, in which acute inflammation attacked the epiphyses of bone, or the diaphysial cartilage; in which by a like treatment good results were secured.

CASE 13. *Acute necrosis of the head of the tibia, with inflammation of the joint; opening of the abscess in the bone, and removal of a sequestrum; rapid recovery.*—Robert G—, aged nine, came under my care at Guy's Hospital on October 4th, 1887, with acute periostitis or ostitis of the head of the left tibia, which had followed closely upon a blow he had received when running seven days previously in a fall upon the knee against a kerbstone. The accident was followed by pain, which became severe on the second day, when swelling appeared. He could not then walk. When admitted on the fifth day of the symptoms, there was effusion into the knee-joint, extreme pain on movement, and tenderness, but no fluctuation over the head of the tibia. His temperature was 102·6°. An incision was at once made down to the bone, but blood alone escaped. This operation, however, gave relief, but only for a time. Consequently a second incision was made on the inner side of the bone, and later on others, for diffused suppuration took place all round the head of the bone, which suggested more than the possibility that an abscess existed in the bone with more or less necrosis. On January 1st the bone was consequently explored by means of an incision and chisel, and a large cavity in the head of the tibia exposed, in which rested a sequestrum. This was then removed, and the cavity well cleansed with sponges dipped in hot iodine water. The bone left seemed to be a mere shell, the *upper surface of which formed the articular facet of the joint.* The cavity in the bone was then allowed to fill with blood, and iodoform dressing was applied over the wound, with a sponge as a pad, and the whole limb was fixed on a posterior splint. All constitutional symptoms rapidly disappeared after this operation. The wound quickly filled up, and in six weeks the boy was practically well. In another three weeks he was allowed to move the knee, when the movements became natural. By the end of March he was quite well, and left the hospital with a perfectly movable knee-joint. In this case an earlier exploration of the bone should doubtless have been made. Such a measure would probably have limited the disease, and would certainly have saved pain.

CASE 14. *Acute abscess, with necrosis in the head of the tibia; free incision into the bone; removal of sequestrum; recovery.*—George H——, aged fifteen, a healthy boy, came under my care in July, 1875, with a tensely swollen and inflamed left leg below the knee, and a discharging sinus over the inner tuberosity of the tibia. The boy was in great pain, and was very ill with high fever. The swelling had been present for three weeks, and had followed a blow below the knee. An abscess over the head of the tibia had been opened two days previously. As there was not sufficient drainage for the abscess, a free incision was at once made over the head of the tibia, and the parts were irrigated. A fistulous opening was then discovered leading into the bone; this was at once enlarged, when a cavity in the head of the bone the size of a walnut and lined with membrane, was exposed. The cavity contained a rounded sequestrum, which was removed, while the cavity of the abscess was cleansed. After this procedure rapid recovery ensued, and the boy left the hospital on August 30th, well, with a sound limb and movable knee-joint.

The following cases, kindly furnished to me by my colleague, Mr. Davies-Colley, further illustrate this point.

CASE 15. *Drilling for acute necrosis of tibia; suppuration of knee-joint; incision into joint; recovery, with movable knee.*—Percy C. C—, aged nearly four, was admitted into Guy's Hospital on March 16th, 1885, with acute inflammation of the right tibia. On the 11th his leg had begun to be painful. His father said that no cause was known, but on a subsequent occasion he told us the boy had had a fall the same day or a day or two before, and that he had some fits before the knee pain began. On admission the boy's temperature was 102·8°. Fluctuation was felt over the whole of the subcutaneous portion of the right tibia. Free incisions were made down to the bone, which was found to be nearly all bare, and was drilled in four places, whence pus welled out. On April 27th the patient's temperature was 105°. On the 30th of that month the right knee-joint was swollen and fluctuating. Free incisions into the joint were made, giving exit to thick curdy pus. At the same time the whole of the anterior part of the diaphysis of the tibia was removed. The boy recovered, with a perfectly movable knee-joint.

Case 16. *Acute necrosis of the lower half of the tibia ; suppuration of the ankle-joint ; drilling of bone and incision into joint ; recovery.*—Fred. D—, aged sixteen, was admitted on May 12th, 1882. He had been a little out of sorts ; had fallen and hurt his left ankle. Œdema of the leg and high temperature followed. On May 18th a free incision was made down to the tibia in its lower third ; the ankle-joint, which was suppurating, was also opened. On May 23rd the temperature, which had at first fallen, had again risen ; and the patient had had some vomiting. Some swelling was found just below the tubercle of the tibia. Ether was again administered, and pus evacuated by a free incision. The bone was then drilled, when two or three drops of pus oozed out from the drill holes. From the fact that the second periosteal abscess was separated from the first by apparently healthy bone, Mr. Davies-Colley thought that the disease had extended from one part of the bone to the other by suppuration along the medullary cavity, and that it was desirable to give exit to the pent-up pus. After necrosis of nearly the whole shaft the boy went out quite well, with a stiff ankle. Mr. Davies-Colley remarks : " If I had drilled when I cut down upon the lower third of the limb, I might have prevented the extensive necrosis. So I determined in future cases to drill earlier."

CASE 17. *Subacute inflammation of the tibia ; incision down to the bone ; abscess in the bone ; subsequent recovery.*—Martha F—, aged fourteen, was admitted on July 16th, 1882. She had injured her left ankle a week previously. There was swelling of the ankle and over the lower half of the leg, and semifluctuation over the tibia. A free incision was made, but only serum came away ; the periosteum was not separated. On September 22nd a sinus was still left. An incision was then made down to the bone, which was found to be enlarged, and converted into a cavity containing pus, lymph, and small sequestra, which had to be scraped away. The patient was discharged convalescent on November 21st, 1882.

Of the foregoing case Mr. Davies-Colley remarks :

"If I had drilled at the first operation I should very likely have saved her a long illness and some danger."

If this practice were carried out systematically in all cases of acute inflammation of bone and its covering, I do not believe we should hear so much as we do of subperiosteal resection of the shafts of inflamed bones—an operation I have never done, nor seen to be needed; and amputation for acute endostitis would only be called for in neglected cases, or where joint complication rendered any partial measure impossible. At any rate, it may be safely said that the necessity for either of these severe measures would rarely occur. In any local periosteal inflammation, in which symptoms are less severe than in the cases quoted, the same principle of practice is equally applicable, and, it may be added, beneficial. Thus in my own practice I have, with invariable advantage, in not a few examples of local periostitis, where local pain was a serious symptom, made a free subcutaneous incision down to the bone through the inflamed periosteum by means of a tenotomy knife carefully introduced through a valvular opening in the skin; the incision not only relieved tension, the cause of pain, but at the same time checked inflammatory action. Where this little operation is done early, before suppuration has taken place or the vitality of the bone has been jeopardised, a rapid recovery may be looked for; and where more serious structural changes have taken place, pain is relieved, and, there is reason to believe, the destructive changes are limited. In the local periostitis or ostitis of a cranial bone, with or without fracture, this detail of practice cannot be too speedily carried out; but in these cases an open incision is probably to be preferred. In the following case the benefit of the treatment was well exemplified.

CASE 18.—*Periostitis of both tibiæ after typhoid fever; suppuration of one limb and necrosis; subcutaneous division of the periosteum of the other; recovery.*—In March, 1865, I was consulted by Mr. B—, aged twenty-six, for some great thickening of the shafts of both tibiæ, which had appeared seven weeks previously when he was recovering from typhoid fever. There was much pain in both limbs, and in the right apparently suppuration, but not in the left, which was hard to the touch and very painful. The swelling over the right tibia was laid open by a free incision, and pus evacuated, the bone beneath being evidently necrosed. The swelling on the left limb was divided subcutaneously down to the bone by means of a long tenotomy knife; blood alone escaped. Pain was at once relieved, and did not return; and a rapid recovery took place. In the right limb some dead bone was eventually removed, and recovery followed.

Having thus far considered the effects of tension acting at a high level in acute inflammation of the

periosteum and bone, and dwelt upon the principle
of practice which should be acted upon for its
relief, with the best means of carrying it out, I
propose now to pass on and consider the effects of
tension in the more chronic and subacute affections
of the same structures; and trust that I may be
able to convince you that it is by acting upon the
same practical principle that the best hope can be
entertained of arresting the inflammatory action
and of placing the inflamed part in the most
favorable condition for repair. To allow these
cases to take their course, every surgeon will
admit, is most unsatisfactory; for their course is a
long one, often for years; and they commonly, if
not invariably, end in either local suppuration,
local necrosis, or sclerosis of bone, but rarely in
recovery by resolution; and in any individual case
one or more of these results may be found together.
The preparations before me demonstrate the truth
of these remarks. The object of the surgeon in
the treatment of these cases should therefore be
to anticipate these organic changes, and prevent
them where he can. Attempts to do this by
medicinal and local measures, other than opera-
tive, are neither hopeful nor successful, although
the value of the iodides and of mercury in certain
cases must be recognised; and I feel confident
that the only sound principle of practice which
should be carried out in examples of chronic
ostitis is the one based on the relief of tension,

as indicated by pain. To illustrate my meaning better, let me place before you a typical and not uncommon case. A surgeon has before him a patient, usually a child or young adult, who complains of a dull, aching, stubborn pain, localised in the shaft, or more commonly in the extremity of a long bone. The pain has probably existed off and on for months, or possibly for years, and it is to a degree constant; it tends, as John Hunter expressed it, "rather to produce sickness than to rouse." After exercise, or use of the affected limb, the pain is intensified, sometimes to a high degree, and at night, when the patient is warm in bed, it is to a certainty aggravated. In the sub-acute cases fever or febrile disturbance may exist; in the more chronic cases it is rarely met with. On local examination, there is probably, if the affection has been of some standing, an enlargement of the bone, and on passing the palm of the hand from above downwards over the seat of mischief an increase of temperature is to be detected. Firm local pressure recognises in some cases a tender spot, whilst in others local percussion through the surgeon's finger elicits pain. There may or may not have been a history of some local injury months or years previously. There may or may not be some general symptoms; as a rule, the general symptoms are conspicuous by their absence. In such a case as this the only diagnosis that can be made is one of inflammation of the bone, which has been persistent and progressive; but whether the case be one of osteo-myelitis of the articular extremity of the bone, as one author might like to call it, or of rarefying osteitis, as modern authorities would describe it, is of small importance. In simple pathological language, the case is one of inflammation of the shaft of the bone, or of the cancellous bone situated at the epiphysial extremity of the diaphysis (juxta-epiphysial ostitis), or of the epiphysis itself, which experience tells us in no uncertain tone, if it be allowed to continue, will end in some destructive change, with more or less condensation of bone structure. What change the bone may have undergone at the time when it came under observation must be a dark point, and one quite beyond the power of the surgeon to diagnose with any certainty. There is happily, however, no great necessity for the surgeon before he acts to form any very definite diagnosis upon these points. He need do no more than recognise the presence of inflammation in the bone, and of tension as indicated by pain, to induce him to take steps for its relief; and a very little experience will prove to him that by the relief of tension in the bone, wherever it may be found, pain will be relieved, if not made to disappear, and with its amelioration the inflammatory process will probably subside. Under the circumstances now being considered, I should at once make an incision down to the bone through the periosteum, and perforate the bone with a drill, well into its centre in one or more points, according to the extent of bone involved; and I should do this with the object of relieving tension in the bone. If blood alone or serum escaped from the punctured wound, I should conclude that no destructive changes had taken place in the bone, and consequently I should do no more, in the confidence that the simple operation I had performed would, in all probability, be followed by the immediate and permanent loss of pain—which, it is to be remembered, is the measure of tension—and by the cure of the disease by resolution. Should, however, the local pain persist after this operation, I should drill again; and if this measure proved unsuccessful, I should infer that in all probability some abscess cavity is present which must be sought for. The trephine under these circumstances would be the right instrument to employ, with the drill as a bone-searcher introduced through the trephine excavation. Should pus or puriform fluid escape from the drill opening made in my first exploration, I should have recourse at once to the use of the trephine, cutting forceps, saw, or chisel, with the view, not only of making room for a full exploration of the abscess cavity, but also for the removal of any sequestrum which might be present in the cavity; and last, but not least, the cavity itself should be thoroughly wiped out and cleared of all lymph, caseating pus or old inflammatory material, which if left would do harm. In subacute and chronic ostitis, the single inducement which should lead the surgeon to interfere is deep-seated pain, and particularly nocturnal pain; for the pain not only clearly indicates tension of the inflamed tissues, but also suggests steady progress of the trouble. All experience shows that in these cases, inflammation continues until relief to tension be found either by the destructive processes of the disease itself or by the surgeon's art. The following brief notes of cases well illustrate the practice I have sketched out, with its results.

CASE 19. *Ostitis of the head of the tibia cured by drilling.*—George S—, aged thirty, came into my hands for treatment on April 5th, 1871, with great enlargement of the head of the left tibia. The trouble had appeared about four months previously, with pain, and it had followed a blow. The pain was constant, and of a dull aching kind ; it was far worse at night. The knee-joint was unaffected, and the soft parts over the bone appeared to be natural. There was no history of syphilis. The diagnosis of ostitis having been arrived at, I made an incision down to the bone, and in so doing found the periosteum much thickened. I then drilled the head of the tibia in two places, and from the openings in the bone blood alone escaped. The operation was followed by immediate relief, and pain did not return. The bone gradually diminished in size. The wound healed kindly in a month, and the man left the hospital well in six weeks.

Some months later he was at work with a sound limb.

CASE 20. *Abscess in the shaft of the fibula, drilled and then trephined; cure.*—Frank R—, aged twenty-eight, a healthy commercial traveller, who had never had syphilis, came into Guy's under my care in January, 1885, with a hard, painful swelling over the centre of his right fibula, and some œdema of the foot and parts around. There was locally increased heat. The swelling had been gradually coming on for four months, and it appeared with pain. The pain had been constant since, although it was worse at times, and invariably so at night. On several occasions he had had to lie by, and for three months had gone about on crutches. An incision was made down to the fibula through the peronei muscles, which were found to be infiltrated with inflammatory material. The periosteum over the bone was thickened. I drilled the bone in two places, and from each opening pus exuded. I consequently trephined the bone, and exposed a cavity from which at least an ounce of pus escaped. There was no sequestrum in the cavity. The cavity was washed out with iodine water, cleaned thoroughly with sponge, and dressed with iodoform gauze, &c., and within six weeks the man was well.

CASE 21. *Abscess in the head of the tibia following a blow, which discharged into the popliteal space;*

free opening into the bone; recovery.—Robert J—, aged forty-three, came under my care in July, 1869, with his left leg flexed upon his thigh, a discharging sinus in the popliteal space, and marked expansion of the head of the left tibia. There was constant pain in the bone, and three inches below the knee firm pressure could not be borne. Twenty-five years previously he had had ostitis of the tibia, followed by the exfoliation of bone six months later and recovery. He remained well for twenty years, when he received a blow upon the inner side of his knee, which was followed by a popliteal abscess. For the previous nine months his leg had been gradually becoming flexed upon the thigh. The case was regarded as one of abscess in the upper part of the tibia discharging into the popliteal space. Under these circumstances, the abscess in the bone was opened by means of a strong knife, the bone being thin, and about an inch of the abscess wall in the bone was removed. A large cavity was thus exposed, which extended upwards towards the knee-joint. The cavity was lined with granulations, and contained pus, but no dead bone. It had evidently opened into the popliteal space. The cavity was well washed out, with the popliteal abscess. The knee was extended, and the whole limb fixed upon a posterior splint. The wound was dressed carefully with dry dressing. Subsequently repair went on uninterruptedly. The cavity in the bone filled up; all pain ceased; the popliteal abscess rapidly closed; and by August 12th—that is, five weeks after the operation—the man was convalescent and the wound had healed. A month later the cure was complete.

CASE 22. *Chronic ostitis of the distal ends of the tibia and fibula, after a compound fracture of the bones; great expansion of bones; drilling of bones, with rapid recovery.*—Herbert G—, aged twenty-four, came under my care on June 9th, 1886, with a compound comminuted fracture of the tibia and fibula, about four inches above the ankle-joint. He did well after the accident, and left the hospital on August 6th, with a protective splint, convalescent. Four months later he returned, with great enlargement of the distal ends of the broken bones and much local pain, which was aggravated by exercise or pressure. There was likewise increase of heat in the part. The foot was raised and fixed upon a splint, and cold applied by means of Leiter's coil, with some benefit and relief to pain. He then left the hospital, but returned on February 7th worse than ever—that is, the bones were expanded and more painful, and the pain was worse at night. The general appearance of the ankle suggested the presence of some new growth. On March 4th, 1887, I made a free incision down to the tibia over the enlarged and inflamed bone, and then drilled the bone, when blood alone escaped from the wound. The operation was followed by immediate relief and the steady diminution of the swollen bone. On March 28th the same operation was performed upon the lower end of the fibula, with a like result. The man left the hospital, well, on May 5th, and reported himself three months later as quite well.

CASE 23. *Ostitis and central abscess, following fracture of the shaft of the humerus twenty years before; drilling and trephining; recovery.*—George W——, aged twenty-seven, was admitted into Guy's Hospital under Mr. Davies-Colley, in November, 1885, with pain in the lower part of the arm, thickening of the lower end of the shaft of the humerus, and a small sinus three inches above the olecranon. He had received a simple fracture of the humerus twenty years previously. For six years he had remained well; then, from time to time, there had been a discharge of spicula of dead bone. There was good movement of the elbow-joint. On November 17th, the patient being etherised, Mr. Davies-Colley made an incision down on the back of the lower three inches of the humerus, which were thickened and rough. A small cloaca existed at the upper part of the olecranon fossa. The back of the humerus was drilled three inches up, where the sinus was, and where he had complained of much tenderness. A cylindrical abscess cavity was exposed and Mr. Davies-Colley removed by a trephine, &c., nearly all its posterior wall. It was three inches long by three quarters of an inch in diameter, and communicated with the small cloaca in the olecranon fossa. There was no sequestrum in it. He recovered rapidly.

This clearing of the abscess cavity after its evacuation, or the removal of a sequestrum, is best done by sponge. A piece of sponge on dressing forceps acts as a fine raspatory on a bone cavity, and cleanses it more thoroughly of all inflammatory products and molecular fragments of necrosed bone, than anything else. It is far better than scraping or gouging the bone. The sponge, I have dipped in hot iodine water before use. When the cavity has been well cleared out, and no foreign body in the shape of necrosed bone is left behind, a recovery with a good limb may be expected to result. After the operation described I leave the case much to nature, but take great care to keep the wound aseptic and the parts involved perfectly immovable. If the cavity fills up with blood-clot I do not disturb it, but keep its surface dusted with boracic acid and iodoform powder, and covered with a simple dressing of iodoform gauze dipped in terebene and oil. In the daily dressing, iodine water is employed to irrigate the wound. By these means I have had many cases in which the cavity rapidly filled up with granulation tissue, beneath and within the blood-clot, and subsequently cicatrised without any trouble; the presence of the blood-clot, kept aseptic, doubtless

much helping—in ways I need not stop to consider—
the growth and organisation of granulation tissue.
Some of these cases have been in subjects who were
feeble, and who might have been called tuberculous,
and yet in them repair went on as favorably as in
the apparently more robust. Where with inflammation of the articular extremity of a bone—whether
originating in an epiphysis, diaphysis, or intervening
cartilage —the joint itself is implicated, the case has
of necessity a far more serious aspect. And the
fact should at once lead the surgeon to be more
energetic in his treatment, rather than dilatory ; for
if the joint complication shows itself only as a subacute or chronic synovitis, such a complication may
be expected to subside, if its cause, the bone inflammation, be relieved. In the following case the truth
of these observations is well illustrated.

CASE 24. *Chronic ostitis of the head of the tibia,
with joint symptoms ; drilling of bone ; recovery.*—
Henry R—, a master butcher, aged twenty-eight,
came under my care in October, 1887, with some
enlargement of the head of the right tibia, and a
constant aching pain in the part. The pain had
existed for *ten* years, having originated when he
was eighteen, and had come on without any known
cause. Every few months he had been quite inca-
pacitated. At times the soft parts over the bone
swelled, particularly after violent exercise. The
pain likewise then became much aggravated, pre-

venting sleep; also, it was always worse at night. The knee-joint had been stiff for some months, any attempt to flex it exciting pain. There was no history of syphilis. On admission, there was impaired movement in the knee-joint, and some swelling. There was very evident enlargement of the head of the tibia, and more particularly about its inner tuberosity. The leg at this part measured nearly an inch more in circumference than its fellow. At one spot there was increased tenderness over the bone, and the soft parts covering it were slightly thickened. The diagnosis was chronic inflammation of the bone. I accordingly made an incision over the painful spot, and reflected a thickened periosteum. I then drilled the bone well into its centre. Blood alone escaped. I made only one puncture into the bone, as I was anxious to test its efficacy, but was prepared to do more should relief not be afforded. The wound, which I left as an open one, was dressed with care, and the limb fixed on a splint. Everything went on well. All pain ceased after the operation, never to return; and the wound healed in a month. At the end of that time the patient was allowed to move his knee, which he did without pain; and in another month he was quite well. At the present time (five months after the operation) he is well and at work, somewhat surprised at the rapid relief he obtained for a trouble he had had so long.

Case 25. *Drilling for chronic ostitis ; recovery.*—
G. A. T—, aged seventeen, was admitted to the hospital under the care of Mr. Davies-Colley, on Jan.
30th, 1884. He was a tall, thin youth, who had
suffered pain in the lower end of the left tibia for
two years. No cause for its origin was assigned. He
was at first treated with iodide of potassium and perchloride of mercury, and with the local application of
Ung. Hydrarg. Co., but no improvement resulted.
On April 30th, 1884, the tibia was drilled in eleven
places to the depth of from half an inch to an inch.
The bone was soft. Nothing but blood came away.
On May 5th there was freedom from pain. On May
8th, as there was some pain at night, he was ordered
a mixture of potassium iodide and the perchloride
of mercury. On May 25th the patient was discharged cured. He had no pain and walked quite
well.

Case 26. *Chronic inflammation of the upper third
of the tibia following a blow three months previously ;
drilling ; recovery.*—Henry T—, aged thirty, a
carman, married, and having three children, was
admitted, under Mr. Davies-Colley, on March 22nd,
1876, with the upper part of the right tibia swollen.
There was no redness over it, but some tenderness.
There was severe pain in the bone, which was worse
when he set his foot to the ground. He suffered
also at night from a pain of a "jerking" character,
which woke him up. The swelling had come gradu-

ally after a kick from a horse received three months before his admission. He had never had any venereal disease. He was kept in bed; Ung. Hydrarg. Co. was applied, and iodide of potassium administered at first in three-grain and afterwards in ten-grain doses, three times a day. After three weeks he went out relieved; but he soon returned with the pain as bad as ever. He was again admitted on May 1st, 1876. On the 13th, an incision was made under an anæsthetic down to the tibia; and the bone was drilled with a gimlet. A week later, the wound was nearly healed and the pain had gone. The man left the hospital quite well on June 7th.

When an abscess has made its way into an articulation, the destruction of that joint is a general result, and under these circumstances amputation is often called for. To prevent this complication must ever, therefore, be a surgeon's great aim, and this can only be done either by checking the inflammatory action in the bone before suppuration and destructive changes have taken place, or by finding a vent for the external discharge of the abscess. The experience of every surgeon will supply abundant evidence of the truth of these remarks; and it would be well if we could say from the same experience that an equal number of cases could be quoted showing the advantages of early interference. Some few cases doubtless could be adduced; but I am disposed to think that too many are allowed to drift, and hopeless joint disease ensues, which has probably to be met by some trenchant operation.

CASE 27. *Abscess and necrosis of the head of the tibia after typhoid fever; removal of bone, including a portion of the articular lamella of bone forming the knee-joint; recovery, with a good joint.*—Ada M—, aged sixteen, came under my care in June, 1884, having within three months after recovery from typhoid fever—that is, seven months before admission into Guy's—had pain in the head of the left tibia, associated with swelling and high fever, which culminated in an abscess. This had been opened, and had discharged ever since. When seen there was much effusion into the knee-joint, and a sinus existed over the inner tuberosity of the tibia, which led into a cavity in the bone, and the leg measured at the level of the sinus an inch and a half more than its fellow. The bone was consequently at once trephined and a sequestrum removed, and, whilst the exposed cavity was being rasped, another abscess was opened nearer the joint. The sequestrum removed included a portion of the articular surface of the tibia. The head of the bone formed a mere shell. The cavity was well syringed out with warm iodine water and dressed with iodoform gauze dipped in terebene, and the leg and knee were fixed upon a splint. When the dressings were removed on the second day, the cavity in the bone was found to be filled with blood-clot. Care was taken not to disturb this, but to keep it aseptic by washing its surface daily with iodine water, dusting it with boracic acid and iodoform powder, and dressing it as before. In six weeks the whole cavity had filled up with granulation tissue, the granulations having quietly taken the place of the blood-clot, which gradually disappeared. In another month the external wound had cicatrised, and there was movement in the joint. Six months later the limb was well and sound, with all its movements. The temperature of the patient during treatment never exceeded 99°.

CASE 28. *Acute necrosis of the shaft of the tibia; with suppuration of the knee-joint; perforation of the epiphysis into the articulation; recovery with a stiff joint.*—George H——, aged seven years, came under my care in May, 1879, with redness and great swelling of the upper part of the right leg and œdema of the foot. The child was in intense pain, and cried on the least movement. He had a temperature of 103°, and his pulse was 125. The pain and swelling had existed for three days, and had followed a knock upon his knee against the edge of a wall two days previously. The centre of the swelling and the pain seemed to be over the inner tuberosity of the tibia. The knee-joint was enlarged. I at once made an incision down to the bone at this spot and evacuated pus from beneath the periosteum, and this measure gave relief. The knee-joint, however, suppurated, and had to be incised, irrigated, and drained on June 24th; that is, about six weeks later. During these weeks the acute symptoms subsided, but it was clear that extensive necrosis of the diaphysis of the tibia had taken place; and on August 7th I removed nearly the whole of the dead shaft of the bone. Having done so, I readily found an opening which had passed through the upper epiphysis of the tibia into the knee-joint, thus explaining its suppuration. In about six months the child was well, with a stiff knee.

In hip disease, secondary to bone trouble, its most common cause, the value of the practice of tapping bone is well seen, and in the following two cases well demonstrated. In one, the bone trouble had not passed beyond the hyperæmic stage; in the second, an abscess had formed and was opened. In both, good results ensued.

CASE 29. *Ostitis of the trochanter and inflammation of the hip cured by trephining the bone.*—Walter S—, aged seventeen, who had enjoyed excellent health, came under my care in July, 1875, with symptoms of hip disease. They had appeared seven weeks previously as pain in the left hip when he walked, but with absence of pain when he rested, except in bed, when the pain kept him awake. The pain steadily increased and became more constant, so that at last he could not stand. On admission, the thigh on the level of the great trochanter measured an inch and a half more than its fellow. The great trochanter was much enlarged, as if from its expansion, and pressure upon it excited pain. The soft parts over the bones looked natural, but to the hand felt hotter than those over the corresponding bone. Pain was produced by movement of the bone, and on pressure into the joint behind the trochanter. I looked upon the case as one of chronic inflammation of the great trochanter and neck of the femur, with inflammation of the hip-joint from its contiguity, and I advocated tapping the bone. This I did on August 20th. I first made an incision three inches long over the trochanter down to the bone and peeled off some thickened periosteum. I then trephined the bone, which I found soft and very vascular. Through the trephined orifice I then drilled the bone in four directions; no pus, only blood, escaped. The wound was dressed in my usual way, and the limb fixed in a double splint. No bad symptoms followed this operation. The patient lost his pain from the date specified, his joint symptoms subsided, and the wound healed in three weeks. He left the hospital convalescent, and when seen by me some six months later he was quite well, though his hip was stiff.

CASE 30. *Hip disease, with expansion of the great trochanter from ostitis; drilling of bone, without benefit ; trephining ; opening an abscess in the bone ; cure.*—A man aged forty-five came under my care in October, 1876, for hip disease of seven years' standing. He had, however, been able to get about, although with pain. Six months before coming under my care he became much worse, and was obliged to remain in bed. For some months the hip had enlarged. When admitted to the hospital his right hip was swollen, and the limb was shortened and adducted. The great trochanter was much enlarged. Movement of the limb caused pain, but the head of the femur moved smoothly in the acetabulum. Pressure over the trochanter could not be tolerated. The pain in the hip was worse at night. On October 24th the great trochanter, which was so much enlarged, was cut down upon and drilled in six places. Blood alone escaped. The bone felt soft, and was clearly vascular. No relief, but no harm, followed this operation. Consequently, on the 27th I trephined the bone, and dropped into a cavity containing lymph and pus. This was well cleared out, and its walls gouged. From this time all pain and bad symptoms disappeared, and the wound granulated. On February 28th the limb was abducted from its position of adduction, and fixed on a double splint; and by April the man was well. Three months later he could walk upon the limb without pain, although the movements of the joint were not complete.

The practice inculcated by the last two cases

might, I think, be repeated more frequently with advantage.

Chronic circumscribed abscesses in bone are to be diagnosed or suspected by the presence of the same symptoms as have been given under the heading of chronic inflammation of bone generally. They are, however, often met with after many years of symptoms. In Brodie's two celebrated cases there had clearly been local evidence of bone disease for ten or twelve years respectively, about half the patient's life. These abscesses are, moreover, found in the epiphysial extremity of a diaphysis, or in an epiphysis (a rare position). The pain of a chronic abscess in bone is, in exceptional cases, very slight; more often it is severe; it is, as other bone pain, paroxysmal and worse at night. One patient described it to me, when it was at its worst, as being "like the falling of drops of molten lead." Others describe it as a toothache pain or throbbing pain, or like the ticking of a clock. With these symptoms, abscess may with confidence be diagnosed. An abscess may appear as a direct result of inflammation, but as often as not it is consecutive to some antecedent action, to some ostitis which had been treated and regarded as cured. It is, under these circumstances, due to the breaking down—caseation—of some old inflammatory products which nature's processes had failed to absorb or utilize; and it should therefore be regarded as a residual abscess in a bone, since it is analogous to such as are found in the soft parts, and particularly about joints.

Brodie, without doubt, was the first surgical author who, by his papers of 1845* and 1846,† caught the ear of the profession of his time, and led his contemporaries as well as those who followed, to look upon such cases as a novelty, and the practice indicated as an innovation. So much so was this the case that Sir W. Ferguson, in 1864, in his lectures from this chair, described "the memorable instance in which Brodie amputated a leg for incurable pain in the tibia as one of the beacon lights of surgery never to be forgotten. It was, if I mistake not," he added, "the model case on which all our modern ideas about abscess of bone are founded, and the pathological examination of that limb led to a line of practice of inestimable value, which even at the present day (1864) is, I imagine, scarcely appreciated at its full worth."‡ Yet on looking into the subject, William Broomfield, a surgeon to St. George's Hospital, where Brodie was educated, wrote in 1773 : "Whenever a patient complains of a dull, heavy pain, deeply situated in the bone, possibly consequent to a violent blow received on the part some time before, and though, at the time the patient complains of this uneasiness within the bone, the integuments shall

---

* 'Lond. Med. Gaz.,' Dec. 12th, 1845, p. 1399.

† 'Med.-Chir. Transac.,' 1846.

‡ 'Lectures on Progress of Surgery,' p. 35.

appear perfectly sound, and the bone itself not in the least injured, *we have great reason to suspect an abscess in the bone.*"* And John Hunter himself, who lectured on surgery at St. George's Hospital, when speaking of abscess in bone in 1787, said in one of his lectures: "The crown of the trephine is often necessary to be employed in order to get at the seat of abscess;" and he spoke as if the practice Brodie advocated later on was well recognised in his day—as, doubtless, it was. Brodie, at any rate, had forgotten it, and he did good in recalling the attention of surgeons to the subject. Brodie, however, went a step further than any other surgeon of his day, for he not only recognised the value of using the trephine when an abscess existed in bone, but he suggested the possibility of the practice being of use for the very purpose I am now advocating— the relief of tension; and he did so in the following words: "Even if abscess should not exist, I can conceive that the perforation of the bone, *by relieving tension and giving exit to serum collected in the cancellous structure, might be productive of benefit;* and at all events, the operation is simple, easily performed, and cannot itself be regarded as in any degree dangerous."† I am pleased, therefore, to be able to bring forward such a surgeon as Brodie to support me in the practice I am now advocating.

* 'Chirurgical Observations.'
† Brodie on ' Joints,' third edition, 1850, p. 298.

CASE 31.—*Chronic abscess in the lower end of the ulna, of eight years' duration; trephining; removal of a small sequestrum from the abscess cavity; complete recovery.*—Emma R——, aged twenty-five, a healthy married woman, came under my care in November, 1881, with an enlargement of the lower two inches of her right ulna, which had been slowly coming for eight years. The affection had followed a sprain. Pain, of a dull, aching character, had followed the sprain and this had at times been felt up the arm. When seen, the bone in its lower two inches was at least twice its normal size. The swelling was smooth, and to the touch painless; all the movements of the hand were natural. There was a constant dull aching pain in the part, and at times the pain was intense. The diagnosis made was that of chronic inflammation. In December I cut down upon the bone, and peeled back the periosteum, which was healthy. I then trephined the ulna, and found great difficulty in doing so on account of the density of the bone. Having removed a cylinder of bone, I exposed a cavity about three quarters of an inch in diameter, in which rested a small round sequestrum of bone, surrounded with pus. The wall of the cavity was made up of dense bone. After the operation all pain ceased, and a good recovery followed. When seen by me five years later the hand and arm were as good as the other, except for the scar.

CASE 32. *Abscess in the shaft of the femur simulating a new growth ; opening of abscess ; cure.*—William B—, a gardener, aged twenty, came up to me from Folkestone, by the advice of Messrs. Eastes of that place, for a swelling occupying the junction of the middle and lower thirds of the left femur, which had been coming on, or at least had been observed, for five months. For two years he had had pain off and on in his left thigh, but he did not think there was any swelling. Two months before being seen he had had an attack of bronchitis, and on his recovery from this the pain in his left thigh became worse, and a swelling was discovered. For two weeks he went about his work, although with pain, which was not, however, very severe during the day, but at night it prevented sleep. For two months he had been unable to get about, and on account of this pain he sought advice. When I saw him on January 6th, 1888, I found an ovoid spindle-shaped enlargement of his left thigh-bone, extending downwards to the knee, and some effusion into the knee-joint. The swelling was very clearly defined, and on gentle manipulation was not painful ; in certain parts, however, firm pressure was resented. The limb measured an inch and a quarter more on the affected than on the sound side. The man went into Guy's Hospital for treatment. Two days after I first saw him a change had taken place in the tumour. It was less defined; the old pain had gone; but the swelling was tender. There was

likewise more swelling in the adjacent soft parts, and
also there was more heat. On January 13th the
swelling was explored through an incision made on
the anterior and outer part of the thigh, over the
tender spot; and when the bone was reached a small
discharging orifice, just large enough to admit a
probe, communicating with the interior of the bone
was found. From this orifice pus exuded. The
abscess cavity in the bone was then laid open by
means of a chisel and strong scalpel, and about two
ounces of pus were evacuated. There was no
necrosed bone found. The cavity in the bone was
smooth. I well washed and syringed out this cavity
with hot iodine water; with the same lotion thoroughly irrigated all the soft parts into which pus had
escaped, and dressed the wound with iodoform gauze
dipped in terebene oil, after having introduced a drainage-tube into the abscess cavity. The limb was then
secured upon a back splint. In the subsequent
history of the case there was nothing striking to
relate, for recovery went on steadily and painlessly,
and in two months the wound had healed. Later on
the wound reopened, and some dead bone, which
had formed the wall of the abscess, was removed.

Trephining a bone the seat of chronic ostitis, or of
what is often called *condensing* ostitis, is an operation of recognised value, although it has, as a rule,
been undertaken from a mistaken diagnosis, and
under the belief that a chronic abscess in the bone

existed. The experience thus gained, has, however, taught surgeons an important lesson, and that relief to a prolonged, persistent, and disabling pain in a chronically inflamed bone may be expected to follow the operation of trephining. Under such circumstances it may well be a matter of surprise that the principle upon which the operation proved of value has not been more fully recognised, and that the expediency of relieving tension in a bone the seat of a chronic inflammation has not been accepted. For it might have been fairly argued that if in a chronically inflamed bone relief to pain may be expected to follow the operation of trephining, even when the disease has persisted for many years, the same operation, or its equivalent, if applied earlier, would not only bring about equally good results, but at the same time tend to prevent or modify the changes in the bone which are generally met with in such cases, and which the preparations in our museums so abundantly illustrate. That there is truth in this argument it is quite impossible for me to doubt; and I am convinced that if surgeons would, as a principle of practice, apply their art to relieve tension in all bones the seat of inflammation, much suffering would be saved and fewer bones and limbs sacrificed; the cases of inflammation of the bone would be far more amenable to treatment, and a principle of practice would be introduced which would bring these hitherto difficult cases of disease within the limits of curable affections, without the frightful destructive changes with which we are all too familiar. By way of illustration I append the brief notes of a few cases in which success followed treatment.

CASE 33. *Chronic ostitis of the shaft of the tibia of sixteen years' standing ; trephining ; cure.*—C. H—, aged twenty-nine, came under my care in January, 1885, with an enlargement of the shaft of the right tibia at the junction of the upper and middle thirds of the bone, which had been coming on for sixteen years, or more than half the man's life, and had followed a severe blow. The bone measured transversely at the spot indicated one inch more than its fellow, and vertically the swelling occupied about three inches of bone. To the hand the affected bone felt hotter than the other, but there was no pain on pressure. There had been during the past sixteen years a constant aching pain in the part, and after much exercise a throbbing. The pain was always worse at night. There was no history of syphilis, and the patient generally was fairly well. Having diagnosed chronic ostitis, I trephined the bone in the centre of the swelling, and used a large trephine. I perforated down to the medullary cavity, although with difficulty, as the bone was very dense. I also drilled the bone in two other directions from the trephine opening. The operation gave full, immediate, and permanent relief to all the symptoms, and a good recovery followed. I saw this gentleman a year later, when he was quite well, the bone having much contracted.

CASE 34. *Chronic ostitis of the shaft of the tibia of four years' duration; trephining; cure.*—Alice J—, aged sixteen, came under my care in December, 1878, with considerable enlargement of the upper third of the right tibia, thickening of the soft tissues over it, and pain of a severe character in the part. The pain, which was at times aggravated, was said to resemble the " ticking of a clock." Her temperature was 100°. The swelling had been slowly coming on for four years after a fall upon a stone. I cut down upon the part, and, having reflected some thickened periosteum, trephined the bone down to the medullary canal. The bone which was removed by the trephine was dense, like ivory. Everything went on well after the operation, and the original pain at once ceased. The wound granulated up in a month, and when seen six months later the patient was well. She had not had any return of pain, and the thickened bone was fining down. The temperature, which was 100° at the operation, fell to normal, where it remained.

CASE 35. *Ostitis of the shaft of the tibia; trephining of the bone, with relief to pain; subsequent necrosis; recovery.*—James W—, aged twenty, came under my care in June, 1860, with enlargement of the shaft of his right tibia from inflammation following a blow eight months previously. The bone was much expanded, and the seat of a constant, and at times intense pain, which was always worse at night. I trephined the bone, and removed a central circular piece of bloodless, dense, waxy-looking bone ; and, as a result, all pain permanently ceased. Some limited necrosis, however, took place at a later period, and after the sequestrum was removed recovery followed.

CASE 36. *Ostitis of the lower half of the shaft of the right humerus ; drilling of bone ; recovery.*—I was consulted by M. R —, a healthy-looking youth, aged seventeen, in September, 1871, for a considerable enlargement of the lower half of his right humerus, which had been steadily increasing for at least five years. It was the seat of a constant pain, which was always worse at night ; at times he had attacks of very severe pain in the part. He could give no history of injury. When I saw him the bone seemed to be uniformly enlarged and smooth. It was not painful on gentle manipulation but firm pressure over the bone excited pain, as also did forcible movements of the joint. I looked upon the case as one of chronic ostitis, and drilled the bone at its outer aspect in three or four places. Blood alone escaped. All pain, however, left the part after the operation and the wound healed. Six months later the patient was well, and the bone had much diminished in size.

CASE 37. *Chronic ostitis of the tibia of at least ten years' duration; trephining of ivory-like bone; discovery of a small sequestrum in the bone; rapid recovery.*—Mrs. M—, a lady of about thirty-five years of age, consulted me in July, 1880, with a tibia which was clearly much enlarged about its centre, and the seat of a dull aching pain for at least ten years. She was married, and had been much in India. She was apparently in other respects healthy. The pain in the bone was constant; but at times, and particularly at night, it was much worse. The bone where it was affected was about twice its normal size. There was no history of injury. I regarded the case as one of chronic ostitis, and accordingly trephined the bone, Dr. Wyman, of Putney, kindly assisting. I used a trephine about three quarters of an inch across, and never had so much difficulty in removing a cylinder of bone, that which was cut through being like ivory. Having cut well into the bone, I removed the cylinder and found at its base a small spiculum half an inch long, resting in a cavity which just fitted it. This sequestrum I removed. The operation at once gave relief, but the wound was some four months in healing. The pain the patient had endured, however, never returned. This lady is now quite well, and has never had any more trouble with, or pain in, her leg.

My colleagues have likewise followed the practice

I am now inculcating. Thus, three cases of acute
ostitis have been treated by drilling; and in all pus
was evacuated. In one of these two cases recovery at
once took place, and in the remaining the necrosis
that followed was only superficial. Twelve cases
of the drilling of bones for chronic ostitis have likewise taken place in the practice of other colleagues,
and in each one relief to pain was at once afforded.
In seven of these cases a speedy recovery followed
the operation, although the symptoms of local ostitis
had existed for years, varying from two up to twenty-
six. In six of the cases there had been severe local
pain from six to ten years. Four other cases were
much relieved by the measure, but it is not known
whether they were cured. In the twelfth case the
hip disease for which the operation was undertaken
was not arrested. The operation of trephining bone
was performed on seventeen other occasions, and in
thirteen of these speedy recovery was the result of
the measure, although in one with subsequent necrosis. In three instances relief was given by the
operation, but the joint disease for which it was
undertaken, with the view to its arrest, continued,
and had to be treated. With these hard facts, based
on an analysis of thirty-two cases, it can hardly be
said that the operation of drilling or trephining
inflamed bone is not successful.

Time tells me that I must now draw to a conclusion; and as I have applied the principle of practice I am advocating to every variety of inflam-

mation of bone, I may be allowed to summarise the whole in the following conclusions :

1. The pain associated with every form of inflammation of the bone or of its periosteal covering is due to tension, and the severity of the pain is a fair measure of its intensity.

2. In acute inflammation of the bone or its periosteum, tension is the chief cause of necrosis; and in the subacute and chronic forms it is a potent cause of their chronicity, as well as of the destructive changes which as a rule follow.

3. The relief of tension, wherever met with, when the result of inflammation, is an important principle of practice which should be always followed. In bone the principle is most imperative, on account of the difficulties under which natural processes act in that direction, by reason of the absence of elasticity or yielding in bone, and by reason of the anatomical arrangements of its vessels which favour blood stasis.

4. To relieve tension in the softer tissues of the body, the local application of leeches, local or general venesection, acupuncture, aspiration, punctures, and incisions may be requisite; whereas, to carry out the same practice in endostitis or periostitis, subcutaneous or open incisions down to the bone, and the drilling, trephining, or laying open of bone by a saw, may be required, the choice of method having to be determined by the requirements of the individual case.

5. In the early or hyperæmic stage of inflammation of bone, before destructive changes have
taken place, experience seems clearly to indicate
that the relief of tension—as indicated by a dull
aching pain, &c.—by means of drilling or trephining
into bone, may arrest the progress of the disease,
and help towards a cure by resolution; whereas, in
the exceptional cases in which this good result does
not take place, suffering is saved and destructive
changes are limited.

6. In articular ostitis, of every kind and variety
and in every stage, this mode of treatment cannot
be too strongly advocated, as tending towards the
prevention of joint disease.

7. In acute or chronic abscess of bone, diaphysial
or epiphysial, the abscess cavity must be opened,
drained, and dressed in the most appropriate way—
the principles of treatment being the same in hard
or soft tissues, although they are modified by the
anatomical conditions of the parts.

With these conclusions, Mr. President and gentlemen, I close my lecture. I fear much that to some
I may have proved wearisome; and more, that I
have failed to convince others of the value of the
practical principle I have brought before you. My
wish has been to carry conviction to every mind;
and where I may have failed the fault must have
been more in the way the subject has been placed
before you than in the matter itself. At any rate,
gentlemen, I have to thank you for your kind atten-

tion, and to express the hope that your own reflec-
tions on the subject will fill up my deficiencies, and
that in the end all will be convinced that the relief
of tension in inflammatory affections generally, if
not always, is as a principle of practice worthy of
adoption.

LECTURE III.

ON CRANIAL AND INTRACRANIAL INJURIES.

MR. PRESIDENT AND GENTLEMEN,—

The subject of tension, to which my two former
lectures were devoted, leads in no unnatural way up
to the consideration of cranial and intracranial
injuries, for most, if not all, surgeons will be ready
to admit that in such cases the evil effects of tension are notably manifested, and that operations
undertaken with the object of evacuating either
effused blood or inflammatory fluids pent up within
a closed cavity such as the skull—that is, for the
relief of tension—are of especial service. It is not
my intention, however, to dwell at any length upon
this aspect of the question, for I have thought it
well to use my present opportunity to bring before
you some general considerations on the subject of
cranial injuries, which I have good reason to believe
are neither sufficiently recognised nor generally
taught,—and more particularly since these considerations have no unimportant influence on surgical
treatment.

As a preliminary reflection, I would emphasise a
point which the surgeon should ever bear in mind

in all head injuries—viz. that the effects of a force applied to the skull are much influenced by its thickness, and that in this matter there is in cranial bones great diversity.   Under these circumstances, a slight blow on a thin skull may bring about a fracture, and no general cerebral injury, since the vibrations originated by the force of impact are expended locally upon the part struck, and are not carried along the bony basal ridges so as to vibrate in the brain structure and bring about mischief. Whereas, whilst a severe external force may fail to cause a fracture of a thick skull, it may start such intense vibrations within the cranium as to cause bruising or laceration of the brain, either of its surface or of its substance, at a point remote from the seat of impact, and even at times produce laceration of the venous sinuses or of the middle meningeal artery.   In the former case the injury on the face of it looks severe ; whereas it may be comparatively trivial, since there is no cerebral injury.   In the latter instance the injury may appear slight, although in reality it is one which bodes death from cerebral mischief.

To say that injuries of the head should always be estimated *primarily* with reference to the amount of damage the cranial contents have sustained, and *secondarily* with reference to the risk of their becoming involved, is to say what all sound surgeons of experience believe and make the basis of their treatment.   And yet students are taught to think

that scalp wounds, fractures of the skull, hæmorrhage beneath the bone, concussion and compression
of the brain, and inflammation of the brain are
separate and independent affections, with diagnostic
symptoms which can be tabulated. The surgeon,
however, who goes to the post-mortem room for in-
formation, knows too well that in every case of
cranial injury of any importance there are certain
brain changes common to most, if not all, which
should be fully recognised and taken into account
in its diagnosis, prognosis, or treatment; and that,
should either a fracture, with or without depressed
bone or intracranial hæmorrhage, coexist in the
same case, such a condition had better be regarded
as a complication of the common factor rather than
as one which stands alone. For example, a man
falls or receives a blow upon the head, and is for a
time " stunned "—that is, rendered more or less
senseless from paralysis of brain function; he is
said to be suffering from " concussion of the brain,"
whatever that term may mean. Another man, as a
result of the same kind of accident, receives in
addition to the " stunning " a scalp wound, with a
fissure either in the vertex or base of his cranium;
and he is described as one who is suffering from a
compound fracture of the skull, either of the vertex,
base, or of both. A third man comes under observation in the same state of so-called concussion, but
with a depressed fracture of bone, complicated or
not with a scalp wound, and as a result of this de-

pression there may or may not be other symptoms, or those present may be intensified. Under either circumstance, however, his case is described as one of depressed fracture of the skull, giving rise to compression of the brain. Yet in all these different classes of cases there is one common injury, one common source of danger, present or remote—viz. the condition of the brain which is associated with the injury, and which has been brought about by the "stunning" force. If the general cerebral injury be trivial, the local complication of a scalp wound, or even of a fissured or depressed fracture, is, although serious, comparatively unimportant; if it be of a grave nature, the local complication must, however great, sink into insignificance.

What, then, it may be asked, is the condition of a brain in a state of so-called concussion? Let us inquire. Concussion of the brain means, in a physiological sense, a sudden and more or less complete arrest of the brain's mental and physical functions, brought about by external violence. The brain in its bony case has been made to vibrate more or less roughly either by some general shake of the whole frame, or by some local violence applied directly or conveyed indirectly to the cranium as a localised or diffused force, the effects of this force upon the brain being of necessity proportional to its concentration and intensity, and in a degree to the age and healthiness of the brain structure and the thickness of the cranial wall. Thus, a concentrated blow

with a blunt or edged instrument, probably, and in
a thin skull certainly, spends its force in producing
a local cranial or cerebral injury; whereas, any
force of a diffused nature, directly or indirectly
applied, and whether causing a fracture of the
cranium or not, more likely brings about some
structural change in the cerebral tissue, remote
from, rather than at, the seat of impact. And
should the brain or its vessels, in either case, have
undergone senile or any morbid change, it is more
prone to suffer seriously from such external violence
than a healthy organ. In every case, therefore, of
injury to the head, the brain is made to vibrate
more or less forcibly; when the vibrations are
feeble, the injury to the brain structure resulting
therefrom is but slight; when they are severe, the
mischief may be great. The complication of a
fissure or fracture of the skull does not of itself, of
necessity, tend to aggravate the cerebral mischief;
although its presence may be regarded as a measure
of the force which has brought it about. In all
cases of cranial injury, therefore, the conclusion is
clear that the cerebral mischief is the common
factor, and the one important point to be taken into
account.

This conclusion subsequently leads to the important question, What are the changes found in the
brain after so-called concussion of its substance, or
rather shaking of its structure? What are the
structural changes, if any, which can be made out

on the post-mortem table? For, of course, an answer to these questions can only be given from the observation of cases which have proved fatal, either directly from the injury, or at some remote period after the injury from other causes. Happily, the answer to this question is neither difficult nor uncertain. For I may say that at Guy's Hospital for at least a quarter of a century there was no case of head injury examined—and such includes every case—in which there was not some coarse brain lesion found, readily visible to the naked eye; in which there was not some contusion of the brain surface, laceration of the brain either upon the surface or within its substance; or more or less hæmorrhage upon or into the brain. In fact, concussion, in a pathological sense, has been, in my experience, synonymous with contusion or laceration of the brain. Other surgeons have expressed the same opinion from this chair. Sir Prescott Hewett, thirty years ago, said: "In every case in which I have seen death occur shortly after, and in consequence of, an injury to the head, I have invariably found ample evidence of the damage done to the cranial contents." Mr. Hilton, who followed him, wrote: "We ought to consider a brain which has been subjected to concussion as a bruised brain." And Mr. Le Gros Clark, who lectured later, stated: "I have never made or witnessed a post-mortem after speedy death from a blow on the head where there was not palpable physical lesion of the brain."

Neudorfer, of the Austrian army, declares that he
has never seen concussion, as so-called, since in all
cases he has examined cerebral injury was found to
exist. And Fano, a celebrated French surgeon, has
also come to the conclusion "that the symptoms
generally attributed to concussion are due, not to
the concussion itself, but to contusion of the brain
or to extravasation of blood." In fact, all authorities now agree that, when death follows a severe
shaking or concussion of the brain, contusion,
bruising, or laceration of the brain is invariably
present, and that when this is not found, the death
is probably to be ascribed to some other than a
cerebral cause; and I shall be able, later on, to
show that when death does not take place as an
early result of the damage, and the patient either
dies of some other affection or of some remote consequence of the injury, the same evidence of cerebral
contusion is generally present. When extravasation of blood upon or into the substance of the brain
follows "concussion" or rather vibrating injury, it
is to be explained in the same way—that is, by some
injury done to the vessels of the brain itself, or to
the venous sinuses within its membranes. When
due to the bruising of the brain itself, the seat of
injury is probably found on the side of the brain
opposite to that of the cranium which received the
blow; the bruising being brought about by what is
rightly termed "contre-coup." A fall or blow upon
the occiput is, as a rule, followed by some bruising

of the anterior cerebral lobes; one upon the frontal region, by a bruising of the posterior parts of the brain; whilst lateral or vertical blows are felt more by the middle lobes. In severe vertical blows the base of the brain itself is bruised. The amount of extravasated blood depends upon the degree of force applied and the healthiness of the vessels in the injured part, diseased vessels easily giving way under a vibrating force to which the healthy would not yield. When the extravasation of blood is upon the surface of the brain, it is either within the cavity of the arachnoid or the meshes of the pia mater; and under each condition the blood gravitates to the base. When the extravasation of blood takes place into the structure of the brain, it may be found in any part of the cerebrum, cerebellum, pons Varolii, or even in the ventricles, the extravasation rarely showing itself in the form of one large clot, but commonly in small and numerous spots of extravasation, which cannot be wiped away, as if from small vessels.

Thus I found in one case of so-called concussion, in which the fatal result took place sixty hours after the injury, from changes brought about by the severe shaking of the brain, unassociated with fracture, that the brain was bruised all over, and blood was effused at the injured spots; the fluid in the ventricles was blood-stained, and the ventricles themselves ecchymosed.

In another case of death from " concussion,"

without fracture, the result of a fall, in a man aged thirty-one, on the fifteenth day after the injury, in whom convulsions and coma supervened, a layer of blood was found universally diffused over both hemispheres, dipping between the convolutions, and passing downwards towards the base. The clot, which was shreddy and of a dull reddish-black colour, had evidently been effused for some days. The surface of the brain beneath the seat of injury was softened ; and at the base, where it had been damaged by contre-coup, similar changes had taken place. The vessels were healthy.

In a third case, where death followed from " concussion," and the vessels were diseased, multiple extravasations were detected after death throughout the substance of the brain.

A fourth case was that of a man (C. K—), aged sixty-five, who came under my care with a scalp wound over the left half of the occipital bone and noisy delirium, having just previously fallen out of a truck on to his head. He had no paralysis or special head symptoms. The pupils were natural ; the pulse was 70, and the temperature 99·2°. He became rational in twenty-four hours, but only remained so for a day, when noisy delirium returned, with refusal to take food. This condition lasted for fourteen days, when he sank, although food was regularly and carefully given by the œsophageal tube. After his death on the sixteenth day from the fall, the anterior lobes of the brain, with the fore parts of the middle lobes, were found much bruised and covered with extravasated blood.

And in a fifth case—that of a man aged thirty-five, who came to Guy's on March 31st, and died on April 1st, 1882, about fifteen hours after the accident—blood was found extravasated over the whole surface of the brain beneath the membranes, and the brain itself was much ecchymosed in fine points. The third and fourth ventricles were full of blood-clot. There was laceration of the fornix and right optic thalamus, and no other lesion. The man had fallen on his head when jumping off a van, and given himself a scalp wound, exposing the bone, to the left of the occipital protuberance. He walked into Guy's, and complained only of headache and a feeling of sickness. He refused to stay in the hospital, and left. Having walked 200 yards, he vomited, and his friends gave him a " small soda." He then vomited again, and being unable to stand he was brought back to Guy's, where he was admitted insensible and comatose, and died, as already stated, fifteen hours after his fall.

When fissured fractures of the cranium complicate brain injuries, and these fractures are the result of some diffused force, the cerebral mischief is not likely to differ from that which has been just described, although, as the force to produce a fracture may presumably be greater than that which fails to do so, the intracranial injuries may be greater from the cerebral vibration. In some cases the brain itself may, in addition, be bruised at the seat of impact. On the other hand, where the force which produced the fracture is concentrated or the brain-case thin, there may be more of local brain injury at the seat of fracture, and less of distant mischief from brain vibration. And with the fracture there may be certain special complications, such as depression of bone with or without compression of the brain, injury to the dura mater, membranes, or brain from the fractured bone or external force, and extravasation of blood between the dura mater and the bone from rupture of the middle meningeal artery or some venous sinus. But all these are complications of, and additions to, the general injury. It must not be forgotten, however, that in exceptional cases a fracture of the skull may take place from a concentrated local violence without producing any cerebral disturbance, particularly over the frontal region.

To prove still more conclusively that " concussion " of the brain means bruising or laceration of the brain, with more or less hæmorrhage, I propose to bring before you the particulars of some cases which have been examined at Guy's Hospital after death, at more or less remote periods after the receipt of the head injury ; and which died from either an independent affection or some remote results of the injury. Such cases are not numerous, but they are valuable, at any rate, for my present purpose. I trust I shall not, therefore, prove wearisome in reading brief abstracts of some of them.

CASE 1. *Cerebral injury; marked evidence of old bruising of the brain found.*—A powerful middle-aged man was found by a policeman in the Borough Market sitting down in a fainting condition. The policeman gave the patient some brandy-and-water, and brought him to Guy's, where he soon became comatose and died. One of his friends said that he had received an injury to his head some weeks (?) previously. After death no external signs of injury could be made out, and there was no fracture of the skull. The anterior and middle lobes of the left hemisphere of the brain were adherent to the base of the skull. The base of the anterior lobe showed brown discolouration from old bruising, and the middle lobe was so firmly adherent to the bone that it tore away. " These changes," wrote Moxon, who made the examination, "could not have been less than several weeks old, probably at least three months." There was a large effusion of recent blood-clot (two ounces) over the right side of the brain. There was no blood in the brain, and no lesion of its membranes. The viscera were healthy.

CASE 2. *Cerebral injury complicated with fractured base of cranium thirty-eight days before death ; bruising of the brain.*—David E—, aged nineteen, came under my care on June 24th, 1874, having fallen off an omnibus on to his head. He was admitted into the hospital, partly unconscious, with profuse bleeding from his *left* ear, which lasted for two days, and

was followed for another eight days by the escape of
a clear fluid (cerebro-spinal). On the second day
after the accident the facial nerve became paralysed.
The man died on the thirty-eighth day from broncho-
pneumonia, having lost all his brain symptoms,
except the facial paralysis. On examination of the
body, a fracture was found in the skull, across the
petrous portion of the left temporal bone, from the
turn of the lateral sinus groove behind into the
middle fossa, laying open the tympanic cavity.
There was no trace of repair in the fracture. The
olfactory bulb on the *left* side was gone, and the
brain about it was bruised. The central parts of
the brain were soft.

CASE 3. *Cerebral injury complicated with fracture
of the base of the skull twelve weeks before death;
evidence of bruised brain.* (Preparation in Guy's
Museum, 1084[55].)—William B—, aged forty-six,
came under the care of Dr. Hilton Fagge in
1873, with vomiting after food and pain over the
pylorus, which proved to be due to abdominal
cancer. He had been an epileptic for four years,
and during that time had been growing darker.
Five weeks before admission, after a fall on the
head from a ladder, he became insensible for a brief
period, and blood oozed from his *right* ear, mixed
with some watery fluid, which continued to flow for
two days. All head symptoms disappeared, although
he occasionally lost himself during his illness. He

died from abdominal cancer. After death, a fracture of the base of the skull, crossing the petrous portion of the right temporal bone, and running the whole length of the meatus auditorius externus, was found. The under surface of the two anterior lobes and of the points of the left middle lobe of the brain presented a tawny yellow discolouration, clearly the result of blood effusion at the time of the injury, and, as usual, it was more marked on the side opposite to that of the fracture. There was also a little superficial softening of the cineritious substance.

CASE 4. *Cerebral injury six weeks before death, followed by abscess.*—William D—, aged thirty, was admitted into Guy's with pleuro-pneumonia, in a condition which soon ended in death, in March, 1873. Six weeks before his admission a beam had fallen upon his head, hurting him, but not producing any marked head symptoms or any bleeding from the ears or nose. Indeed he had not given up work, but continued as an engineer's labourer for four weeks—that is, up to two weeks before admission, when he felt ill, shivered, and had severe headache; he also soon lost his taste for sweet things. When admitted he had no paralysis, only headache and drowsiness. His temperature was normal. He had no convulsions. After death the brain was found, on removal of the skull-cap, to be flattened, so that it appeared to have no convolutions. Its surface was discoloured in parts from what proved to be abscesses. A large and rather old abscess had burst into the hinder part of the right lateral ventricle, filling it with pus.

CASE 5. *Cerebral injury complicated with fractured skull and spine, &c.; death ninety-one days later.* (Preparation in Guy's Museum, 1084[52].) —George G—, aged twenty-seven, came into Guy's Hospital, under the care of Mr. Forster, in July, 1882, having fallen or been thrown out of a third-floor window. He was conscious, and complained of pain in his back. He had a scalp wound over the occiput, and evidently a fracture of the occipital bone. His lower limbs were paralysed from a fractured spine. The patient died of pleurisy. At the necrosy a fissure was found in the right half of the occipital bone, which extended vertically across the root of the petrous portion of the right temporal bone towards the lesser wing of the sphenoid. In the posterior cerebellar fossa there were two offsets, one of which ran down to the foramen magnum. There were but feeble signs of repair in the fracture. The brain at the right anterior lobe was much bruised. There was blood between the bone and the dura mater, and on the inside of the dura mater there was much brown pigment. The eleventh dorsal vertebra was fractured.

CASE 6. *Cerebral injury six months before death, brought about by meningeal apoplexy; marked evidence of old cerebral injury.*—Patrick H—, aged forty-one, was admitted into Guy's Hospital, under Dr. Pavy, in March, 1874, in a comatose condition, and soon died. He had been drinking for some days before, and had had a fit. Six months before, or thereabouts, after a fall down some stone steps, he was brought into the accident ward with a bruise over the right mastoid process and discharge of blood from the right ear. He was partially insensible for ten days, and was stupid for some time after. His pulse ranged from 40 to 50. He left convalescent and returned to his work, at which he continued until he had the fit for which he was readmitted just before his death. At the necropsy no injury of the cranial bones could be discovered. The dura mater over both sides of the brain presented on its inner surface a tawny red colour, apparently stained with old extravasated blood. That covering the left hemisphere was smooth, but that over the right was in part lined with adherent coagulum which in some parts was of a brownish colour, in others black. Some of it was evidently old. This was part of a large mass of coagulum which lay between the dura mater and the brain, on this side flattening the brain. The surface of the brain was so discoloured from staining with blood that the amount of clot was difficult of determination. The brain had evidently been deeply bruised at the time of injury at the inferior surface of both anterior lobes and the summit of the right middle lobe, particularly towards the back of the lateral surface of the right hemisphere, where there was an irregular fissure, with much tawny discolouration of the tissue. The lateral ventricles were healthy. The large effusion of blood clearly came from the rupture of the vessels of the softened part, and was probably in part due to alcoholic stimulants.

CASE 7. *Cerebral injury thirteen months before death; marked evidence of brain injury.*—George L—, aged twenty-two, was admitted into Guy's Hospital under Dr. Habershon in February, 1871, in a semi-torpid condition, having been found by a policeman in a fit after a day or more of drunkenness. He could be roused with difficulty, and when roused moved all his limbs. He had right facial paralysis. His pulse was 44; the pupils were contracted. He gradually sank. He had had a severe injury of his head thirteen months previously, from which it was supposed he had recovered. After death no signs of injury of the cranial bones were discovered. The brain was flattened, and the anterior and middle lobes were adherent to the dura mater at the base, and the brain tore away when an attempt was made to remove it. The membranes of the brain at the base were very thick, and partly opaque white, partly of an ochre-yellow colour; this appearance was confined to the membranes. The aperture into the fourth from the third ventricle was closed by some recent lymph, and the brain bordering the channel was soft. The lateral ventricles were greatly dilated, the descending cornu on the left side projecting like a blister at the base of the brain. The left middle lobe was diffluent.

CASE 8. *Cerebral injury two years before death from phthisis; marked evidence of bruised brain.*—Wm. C—, aged fifty, came into Guy's Hospital with phthisis, under the care of Dr. Moxon, in 1874, and died within a few days. He was a sailor, and had enjoyed good health up to two years previously, when he had a high fall, and was stunned and bled from his left ear. The next day hot fluid came from the ear. He was laid up for three weeks, and when he left his bed he was giddy for nearly one year. He returned to his work, although deaf in his left ear. Five weeks before his admission he " caught a cold and cough," and died of acute phthisis. At the post-mortem examination no clear indication of fracture of the petrous portion of the left temporal bone could be made out, although there was a slight transverse line across the bone. There was an old bruise of the anterior lobe of the brain on the left side and slightly of the base of the middle lobe. That on the anterior formed an irregular hollow the size of a shilling. The grey substance was quite destroyed, and microscopically was found to present hæmatoidin crystals and compound granular masses.

CASE 9. *Cerebral injury complicated with fracture of the base of the skull on the right side eight years before admission, followed by fits, chronic hydro-*

*cephalus, spinal curvature, emphysema, and hydro-
thorax; marked evidence of old bruising of the brain.*
(Preparation in Guy's Hospital Museum, 1084[56].)—
David W—, aged thirty, was healthy until eight
years before admission, when he fell off a ladder and
hurt his head. Since then he had not been able to
do hard work, and his memory had failed him.
Five years ago he had a fit, and others had followed.
He was admitted under Dr. Wilks in 1876, in a fit,
comatose, and passing urine under him; but he
gradually recovered sensibility in a few days. In
the hospital he had a fit, and both sides of his body
were convulsed; when consciousness returned, it
was found that the *left* side of his body had diminished sensation, but he said this had existed for
eight years, and was gradually growing worse. He
was fairly intelligent, and answered questions
readily. His pulse was slow, 38 to 40. He died of
bronchitis. At the necropsy the brain was found to
be flaccid. At the base of the right middle fossa,
and over the roof of the orbit, there was a layer of
brown pigment; also in the posterior fossa. The
same pigment existed in patches over the base of
the brain. There were fourteen ounces of fluid in
the ventricles. The brain substance was healthy.
The foramen magnum was altered in shape, and
narrowed antero-posteriorly, the transverse diameter
being much larger than the other. There was
fracture of the right side of the base of the skull.
The spine was curved at the sixth dorsal vertebra

towards the right. The cord was healthy. The lungs were emphysematous; fluid existed in the left chest.

CASE 10. *Cerebral injury complicated with fracture of the skull nine years before death; evidence of old as well as of recent brain injury.*—Michael L—, aged fifty, came under Mr. Cock's care on December 29th, 1867, and died on January 7th, 1868. He had fallen twenty feet on to the side of his head, and was admitted with a scalp wound and depressed fracture of the left of the vertex. He was unconscious on admission, but not paralysed, and was never clear enough to give a history of his accident. One of his friends stated that he had had an injury to his head nine years previously. After death the cranial bones were found to be thin. On the occipital base, about the left part of the groove for the torcular Herophili, were *fine* rough elevations, and here the dura mater was very adherent. About this part there was an old depressed irregular fracture of the bone, the bone being now united to the rest. Within the skull there was an elevation corresponding to the external depression, and its edges were bevelled off. There was no sign of any old external wound over the fracture, or of blood between the bone and dura mater; but on opening the latter about an ounce of liquid clotted blood was found effused over the left vertex, corresponding to the recent scalp wound. The right anterior lobe of the brain was adherent over the orbit, and the brain here was discoloured brownish yellow, as from old hæmatoma, this part corresponding to the spot of contrecoup from a blow on the left back of the head.

CASE 11. *Cerebral injury complicated with fracture; evidence of brain bruising, recent and old.*—Michael W—, aged fifty-three, came into Guy's Hospital, under Mr. Cock, in February, 1867, and died in two days. He was found in the street insensible, with a wound on the back of his head and bleeding from both ears, but chiefly the left. He died comatose. At the necropsy there was clear evidence of a heavy blow having been given over the right mastoid process, the bone at this part being fractured. There was no blood between the bone and dura mater at this part, but there were about two ounces between the dura mater and the brain. There were *recent* contusions on the anterior and middle cerebral lobes, and *old* yellow discolouration of both middle lobes. The right one was firmly adherent to the middle fossa outside the foramen ovale. Lining the dura mater of the right middle fossa of the base of the skull there was a thin membrane, easily separable; and this was thickest where the brain was adherent. The right middle lobe was more deeply bruised than other parts, where it was fixed to the fossa.

CASE 12. *Compound fracture of the cranium, with marked brain symptoms, occurring three years and a quarter previously; necrosis of the inner table of the skull; removal of the bone by trephining; death from phthisis; evidence of bruised brain and extravasation of blood.*—Conrad H—, aged forty-five, came under my care in January, 1879, with a discharging sinus on the right side of the occipital protuberance, communicating with the interior of the cranium, which had been the result of a compound fracture of the skull he had sustained three and a quarter years previously. The injury was brought about by a blow from a large stone, which stunned him, and he remained unconscious for seventeen days. He was abroad at the time, and had no treatment. About one year afterwards some pieces of necrosed bone were taken from a second fracture on the vertex of the skull he had received at the same time. He had then some slight weakness of his left side. For the last year the sinus had discharged freely, and he had had much headache. Finding on examination that some necrosed bone could be felt within the cranium in the direction of the internal occipital protuberance, I trephined the bone with an inch trephine, and enlarged the opening with Hoffmann's forceps. Having done this, I removed numerous fragments of the inner table of the skull, corresponding with the lateral and longitudinal sinuses. The dura mater within was thickened, and covered with granulations. The operation gave much relief to his head symptoms; but his lung trouble slowly extended, and destroyed life five months subsequently. After death the repair of the base of the skull was found to be complete, but the bone was thick and irregular. The opening in the cerebellar fossa was only partially closed. The dura mater corresponding thereto was thick. The torcular Herophili and sinuses were healthy. The dura mater covering both cerebral hemispheres was tawny with old extravasated blood; indeed, a thin membranous film could be stripped off its inner surface. The summits of the middle lobes of the brain at its base showed some tawny erosion, evidently the result of former bruising. The lungs were extensively diseased.

The evidence I have thus laid before you of the pathological conditions of the brains of those who have suffered from what has been so long known as " concussion of the brain " will, I trust, be deemed sufficient to convince such as may be in doubt that they are really examples of cerebral injury; that term meaning cerebral bruising or laceration, with more or less hæmorrhage upon and into the substance of the brain or its ventricles, the amount of injury of every kind varying in degree in each case. In some it may be very slight, and in others severe.

You will likewise have probably observed—what the details of the cases I have brought before you so forcibly demonstrate—the lasting character of the changes which the brain may have undergone as a result of injury, and the evident slowness with

which nature performs in this organ her reparative
work; for, in some of the cases quoted, years had
passed after the injury had been received, and yet
marked evidence of its former existence was still
present.   Indeed, such evidence as I have laid
before you of necessity draws out the question, Is
a bruised brain ever thoroughly repaired; and are
not the changes which the injury may have brought
about fixed and permanent?   It is to be regretted
that an answer to these questions can only be given
in an unfavorable form; and that, whilst we may
be hopeful as to this complete repair and recovery
from a slight injury or bruise, we are bound to
regard graver cases in a more serious light, and to
deal with them accordingly, since what evidence we
possess seems to show that, when any portion of
the brain has been severely or moderately bruised,
it has been permanently injured.   This evidence
also dovetails in with the general experience, which
tells us not only of the presence of physical head
symptoms, but that the mental and moral characters
of men are often permanently altered by a head
injury.

With these facts and conclusions before us, am I,
therefore, wrong in assuming with some confidence
that you will see with me the expediency of com-
bining with the term " concussion " that of " injury,"
and of describing such cases in the future as those
of injury of the brain from concussion?   The term
" concussion " by itself is vague and delusive, whilst

that of " injury " is clear and true, and conveys at
once a meaning the force of which cannot be mis-
understood.   The word " concussion " later on may
be dropped, and the simple term "injury" retained.
With this starting-point, it would naturally follow
that fractures of the skull in all their varieties,
hæmorrhage into the cranium in all its forms, and
compression of the brain, however brought about,
will be regarded as complications of the one common
and essential factor, cerebral injury, and not, as
now, be regarded as separate and individual troubles
to be dealt with independently.   And even scalp
wounds, the result of external violence, would
assume a position in the surgeon's mind they ought
to have, but have not yet attained ; and consequently
receive the attention to which they are entitled, not
so much perhaps on their own individual account as
simple wounds, but as wounds mostly brought about
by direct violence applied to the cranium, and consequently liable to be complicated with some contusion of the cranial bone or intracranial injury.

Up to this time my observations have been confined to the elucidation of the first of the two main
clinical points which have had a common bearing
upon all cranial injuries, and to which I drew your
attention at an early period of this lecture—namely,
" that all injuries of the head should be estimated
primarily with reference to the amount of damage
the cranial contents have sustained ;" and I trust I
have demonstrated with sufficient clearness that a

cerebral injury of some kind is the one common factor. I propose now, therefore, to pass on and consider the treatment of cranial and intracranial injuries, and, with the light which the above conclusion throws upon the whole subject, see what bearing it ought to have upon the second clinical point to which attention has been drawn—viz. " that injuries of the head should be estimated *secondarily* with reference to the risk of the cranial contents becoming involved;" and it should be remembered that this risk is one to which every degree and variety of cranial injury is liable. In even such an apparently simple accident as a contusion of the head, whether with or without a scalp wound, the fear of this secondary danger ought not to be overlooked; indeed, it ought always to be held in view, for I imagine there are but few surgeons who have not been called upon to treat examples of scalp wounds, or patients who, having had cranial blows and being supposed to have been cured, have, on going to work or moving about, or after some indiscretion of diet, complained of headache, restlessness, giddiness, nausea, or even vomiting, with more or less febrile disturbance—all these symptoms being those of cerebral irritation, or the first step of inflammation. Or possibly the patient has complained only of local pain at the seat of injury, and the surgeon on examination finds some swelling of the soft parts over the bone or some change in the appearance of the wound, its healthy granulating surface having

assumed a pale, flabby condition, always suggestive
of an early ostitis or periostitis. " Acceleration or
hardness of pulse," wrote Percival Pott, " restlessness, anxiety, and any degree of fever, after a smart
blow on the head, are always to be suspected and
attended to. . . . When there is a wound, it
will for a time have the same appearance as a simple
wound. But after a few days all these favorable
appearances will vanish ; the sore will lose its
florid complexion and granulated surface, and become
pale, glassy, and flabby ; instead of good matter, it
will discharge only a thin discoloured sanies, and
the pericranium will separate from the bone. The
first appearance of alteration in the wound immediately succeeds the febrile attack, and as the febrile
symptoms increase the sore becomes worse and
worse. . . . Through the whole time, from the
first attack of fever to the last and fatal period, an
attentive observer will remark the gradual alteration of the colour of the bone, if it be bare. At
first it will be found to be whiter and more dry than
the natural one, and as the symptoms increase the
bone inclines more and more to a kind of purulent
hue or whitish colour." These extracts I have
taken from the works of Percival Pott, who first
drew attention to this danger of bone inflammation
as a result of cranial injury a century ago, but
possibly made too much of it. He regarded it as
the chief element of danger in all cases of contusion, scalp wound, or fracture, and not only tre-

phined the skull when ostitis existed, but laid it
down as a rule that "perforation of the skull is
absolutely necessary in seven cases out of ten of
simple undepressed fracture." The operation for
trepanning was called for in these simple cases "to
prevent the effects of inflammation, detachment, and
suppuration of the dura mater, and consequently
the collection of matter between it and the skull."
From this over-estimate of the value and necessity
of the operation of trephining in cases of fracture,
and the comparative rarity of cases of abscess
between the bone and dura mater as a direct result
of contusion, surgeons have been prone to treat too
lightly the risks of a secondary ostitis following
bone injury, whether complicated or not with scalp
wound or even with fracture, and consequently to
neglect what was good in Pott's teaching. As a
result, I feel sure that many lives are lost yearly,
and that many narrow escapes from death occur.
I have the notes of some fatal cases of the kind
before me, which I have taken from the post-mortem
records of Guy's Hospital. In four of these pyæmia
was the immediate cause of death, and such a result
was probably brought about by the inflammation of
the venous channels of the diploë of the injured
bone. I will not weary you by reading all the
details of the cases, but will lay before you their
chief points.

CASE 13. *Scalp wound; ostitis; necrosis; pyæmia.*

—T. G—, a man aged twenty-two, having fallen from a height, came into Guy's Hospital with a scalp wound and a compound fracture of his leg; he died from pyæmia. The scalp wound had apparently healed, and the man with his compound fracture seemed doing well in all ways, when on the tenth day headache and febrile disturbance appeared, followed by swelling in the seat of cranial injury, reopening of the wound, and suppuration of the parts covering the bone, from which the pericranium had loosened. Indications of pyæmia soon appeared, but no general head symptoms, and he died in two weeks. At the necropsy, the anterior half of the parietal bone beneath the scalp wound was of a whitish colour, bare for a space of three by two inches, and apparently dead. The necrosis extended through the bone, and on the surface of the dura mater beneath the bone there was lymph with pus, as also on the arachnoid surface. The brain was healthy. There were pyæmic abscesses about the body.

CASE 14. *Scalp wound; ostitis and necrosis of the injured bone; meningitis; pyæmia.*—Eliza S—, aged forty, came into Guy's Hospital with a large scalp wound over the right parietal region, and exposed bone. The injury was the result of a fall off an omnibus, which for a few minutes stunned her. The next day she complained of headache, and on the fourth day it was incessant. On the ninth day sleeplessness and indications of fever appeared ; the wound also became unhealthy.   On the twenty-first day there were rigors, and on the twenty-fifth some hemiplegia on the left side, for which the operation of trephining of the right parietal bone was performed, and some fetid pus evacuated from beneath the bone.   The dura mater beneath the bone was velvety.   Convulsions soon came on, and death in a week.   At the necropsy, the whole thickness of the right or injured parietal bone was found to be dead, and the dura mater beneath discoloured.   The surface of the brain corresponding to the dead bone was covered with pus.   Pyæmic abscesses were present in the lungs and liver.

CASE 15. *Scalp wound ; ostitis of the external table, fracture of the inner table ; pyæmia.*—Edw. N—, aged eighteen, having received a scalp wound to the right of the vertex of his skull from a falling bucket, came to Guy's Hospital without any head symptoms, and had his head dressed.   The wound healed rapidly without pain or trouble in about ten days, when some local swelling appeared, followed in two days by fever and rigors.   In this condition he was admitted into the hospital, where pain and swelling of the right elbow-joint appeared, and later on chest symptoms, which proved fatal three months after the primary accident.   After death, the outer shell of the parietal bone beneath the scalp wound, to the extent of an inch and a half by one inch, was

necrosed, but it had not been fractured. The inner
surface of the corresponding area of bone was fractured, the fracture lying in the long axis of the oval
necrosed piece. The fracture was a mere fissure,
slightly starred, and its edges were perceptibly
raised. A little stringy lymph hung about the
bone. Over the dura mater corresponding to the
fractured bone there was a yellowish patch of
lymph. The brain was healthy. There were
pyæmic abscesses in the lungs, liver, and elbow.

CASE 16. *Scalp wound ; exposed and inflamed bone ;
death from bronchitis.*—A man, aged twenty-three,
having been jerked from the shaft of a cart, received
a blow and a wound over his left temple which exposed the bone. There were no head symptoms.
The wound was carefully dressed antiseptically, and
for a week everything went on well. At the end of
that time his temperature went up to 104·6°, and
the glands of his neck commenced to enlarge. On
the tenth day he had paralysis of the left facial
nerve, and bronchitis set in, which quickly destroyed
the patient, three weeks after the injury. At the
post-mortem examination, two square inches of the
left or injured temporal bone were exposed. The
bone was dry and discoloured, but there was no
obvious necrosis. The diploë of the bone in the
line of section of the calvaria on the left side was
more vascular than that on the right side. The
brain and dura mater were healthy. The bronchial

tubes were filled with viscid mucus. The other organs were healthy.

CASE 17. *Scalp wound; necrosis of bone; abscess in the brain.*—A male child, aged three, six weeks before death fell on his forehead, causing a wound one inch above the left orbit. No head symptoms were induced by the fall, and for three days all seemed to be well, when he fainted in his mother's arms. He was then brought to Guy's Hospital, and requested to be left; but to this the mother objected. A few days later the child again fainted and vomited, and as the vomiting persisted the mother brought her child into the hospital. At that time, sixteen days after the accident, the wound had healed, and the scar was adherent to the bone. The child vomited daily, and his temperature was just above normal. A week later, without any important change in the symptoms, the child had a fit which lasted an hour. During the fit the right leg was rigid, and the arm with the fingers and thumb became spasmodically flexed. The left limbs also moved spasmodically. Temperature 106°. In two days the child died. After death the frontal bone at the seat of injury was covered with purulent lymph beneath the pericranium. Beneath this the bone was rough and ulcerated. The inner table over a space the size of a sixpence was opaque yellow. When sawn through, a piece of the outer table was loose and necrosed. The dura mater at this spot was adherent to the brain. There was a little pus on its outer surface. The left frontal lobe of the brain beneath this spot was soft, swollen, and fluctuating, and when cut into was found to contain a large abscess of the size of an apricot, which had a distinct lining membrane. There was no red softening.

CASE 18. *Scalp wound; necrosis of bone; abscess of the brain.*—A middle-aged woman three weeks before her admission into Guy's Hospital received a scalp wound from a blow upon the forehead. She continued at her work for two weeks, and it is said without any head symptoms, when she became comatose and died in a week. She was brought to the hospital in the dying comatose stage. At the post-mortem examination a sloughing wound was found to occupy the centre of her forehead, and beneath it bone was exposed. The surface of the bone was yellow and dead, but not depressed. The area of dead bone was separated from the living by a shallow groove. The inner surface of the bone was blackish; the diploë was full of pus. There was chocolate-coloured pus beneath the dura mater of the injured part, and the brain beneath was suppurating, the abscess burrowing backwards.

CASE 19. *Contusion of cranium; Pott's "puffy" tumour; trephining; meningitis; death; bruised brain.*—W. L—aged forty-six, was admitted into Guy's Hospital, under the care of Mr. Howse, on

December 13th, 1882, fifteen days after having been knocked down by a cab and stunned. He was taken home after the accident, unconscious, and remained in that state for three hours. He then vomited and brought up blood. The next day he became drowsy; and four days later he had a convulsion, which was followed by others for three days. He then became delirious and unconscious. When admitted he was in a low typhoid state, constantly muttering and picking at the bedclothes. Temperature 100·4°. There was no paralysis. A swelling was found over the vertex of the skull. This was cut down upon, and blood was seen effused beneath the pericranium, which readily peeled off the bone. The bone was bruised and yellowish. The inch-trephine was used. The dura mater, which bulged into the opening and was covered with lymph, pulsated. Nothing more was done, except that the wound was dressed and cold applied to the head. The delirium, however, continued, and gangrenous pneumonia set in, which proved fatal on January 13th, thirty days after the accident. After death the base of the brain was found to be bruised by contre-coup, and there was diffused meningitis.

The cases I have quoted were all examples of ostitis and necrosis, the result of a contusion of the bone associated with scalp wound. I could give as many more associated with fracture if they were needed, and they would all tell the same tale. In none of the cases were there any symptoms of brain injury after the accident, and in most the symptoms did not appear for a week or ten days or a fortnight afterwards. In all, the mischief which produced death had clearly originated in the bone. In none of the cases had much care been employed to guard against the secondary mischief which took place, and which led on to a fatal issue; and it may reasonably be thought that, if judicious treatment had been applied from the receipt of the accident, no such result would have been recorded. The conclusion is therefore clear, that all scalp wounds which lead down to bone should be dealt with, for at least a fortnight or three weeks, with much care; and that such cases should, if possible, be treated as in- and not out-patients of hospitals for that time. For it is not the wound treatment only which calls for care, but the patient should be kept quiet, and given a simple unstimulating diet. Stimulants of all kinds should be forbidden, and meat allowed in very limited quantities. If at the end of two or three weeks, or thereabouts, no local or general symptoms appear to suggest mischief, the duties of life may be gradually recommenced. But even then a caution should always be given to observe care.

When local symptoms appear, such as have been described, a free incision down to the bone where no wound exists or a free separation of the peri-cranium where there is a scalp wound, always does good. And should any symptom appear or persist which even suggests any intracranial complication, the operation of trephining should at once be had recourse to. "The spontaneous separation of the pericranium, if attended with general disorder of the patient, with chilliness, horripilation, languor, and some degree of fever, appears to me," says Pott, "from all the observations I have been capable of making, to be so sure and certain an indication of mischief underneath, either present or impending, that I shall never hesitate about perforating the bone in such circumstances. . . . When there is just reason for supposing matter to be found under the skull the operation of perforation cannot be performed too soon; it seldom happens that it is done soon enough. The perforation sets the dura mater free from pressure, and gives vent to collected matter, but nothing more. The inflamed state of the parts under the skull, and all the necessary consequences of such inflammation, calls for all our attention fully as much afterwards as before; and although the patient must have perished without the use of the trephine, yet the merely having used it will not preserve him without every other care." The prevention of this fatal trouble is, however, the more important point to emphasise; and for that purpose I bring the subject before you. Its early treatment may be beneficial and successful; its later treatment cannot be said to be so. Let us, therefore, teach the necessity of keeping patients with all but minor scalp wounds, and with those even where the bone is exposed, quiet and unstimulated for some weeks after the receipt of the injury where it can be done, and by so doing give nature an opportunity of repairing the mischief in the bone, which, though unseen, may reasonably be expected to be present after the application of a force sufficient to produce a scalp wound or a more severe injury. When a blow upon the head is known to have produced a fracture, the case is likely to be treated carefully ; whereas, when no such fracture can be made out, and there is little or no external evidence of injury, the same care is not likely to be observed, although in both cases the violence which had been employed may have been equal. Yet in both cases the dangers of cranial ostitis from contusion are about the same. The fact, I am sure, requires to be emphasised, that cranial contusions, whether associated or not with fractures or with wounds, are always matters of serious importance, and as such should be treated from the first. Having dwelt upon the dangers of these cases, and illustrated some fatal results, I propose to quote a few examples of their successful treatment.

CASE 20. *Scalp wound, and subsequent ostitis, treated by trephining; fissured bone discovered; cure.*—Catherine S—, aged twenty-eight, came into Guy's Hospital in July, 1884, under Mr. Howse, with a scalp wound and fissured fracture of the

skull, the result of a blow from a machine.   She had
no head symptoms, either at the time of the accident
or after, and in six weeks she was discharged, supposed to be cured.   A month later she was readmitted for continual headache and local pain in the
seat of the former injury.   She stated that she had
not been able to work since she left the hospital.
The old wound had reopened a week after leaving.
On her readmission bare bone was felt—indeed, a
piece of dead bone was taken away.   In a day or
so the pain in the head had much increased, even to
make the patient scream.   A crucial incision was
then made down to the skull, when a fissured fracture was discovered.   This operation did not give
relief; consequently, she was trephined over the seat
of fracture.   The bone was found to be very dense
and thickened, and its outer surface rough and
pitted.   No diploë existed.   The dura mater was
rough, and shaggy from adherent lymph.   All cerebral symptoms had disappeared on the second day
after the operation, and a rapid cure took place.

CASE 21. *Contusion of the head ; ostitis ; trephining ;
discovery of fissured fracture; cure.*—Harry D—,
aged five, was admitted into Guy's Hospital, under
the care of Mr. Howse, on September 29th, 1880,
having two weeks previously, in a fall, struck the
left side of his head against a kerb-stone.   He was
unconscious after the accident for a few minutes,
when he vomited.   He was kept in bed for two or

three days, and was supposed to be convalescent, when he had headache, and at night some light-headedness. And these symptoms persisted up to his admission to the hospital. At this time, a fortnight after the injury, a fluctuating swelling about an inch in diameter was discovered beneath the seat of injury behind the coronal suture. There was some fever with night delirium, but no paralysis. He was trephined at the seat of swelling, when a fissure in the bone was discovered, and the dura mater was seen covered with lymph. From this time everything went well; headache and delirium disappeared, and convalescence followed.

CASE 22. *Scalp wound; "puffy" swelling; trephining; cure.*—William T—, aged four, was admitted into Guy's Hospital, under Mr. Howse, on June 5th, 1885, having a week previously been struck above the right orbit by a swing, and received a scalp wound down to the bone. He progressed well for two or three days, when the wound began to inflame and he became feverish. In this condition he was admitted. The wound at this time was sloughing, and the bone was exposed. Temperature 101°. A few days later, as no improvement took place, and the child was drowsy, the bone was trephined at a spot near the wound, which had become "puffy." The dura mater, where exposed, was granular, and the bone eroded. All symptoms at once disappeared, and a rapid recovery followed.

I will now pass on to consider the treatment of head injuries in the light of the view I am now advocating, for I am under a strong impression that such a view cannot do otherwise than have an important influence in rendering treatment more simple and intelligible; since, if in every grave, or indeed apparently uncomplicated example, associated with more or less complete paralysis of at least one of the brain functions, such as is indicated by unconsciousness, the surgeon recognises to the full the force of the fact that the brain as a material organ is bruised or otherwise injured, a line of treatment is likely to be at once suggested which can best favour the restoration of the injured part towards health. Amongst the means which would probably find favour, physiological and mechanical rest would stand foremost, with the administration of nourishment simple enough to maintain the normal powers and help repair, and not stimulating enough to excite action. Everything in the form of alcoholic stimulants or solid meats would be forbidden; and this line of treatment would, moreover, be maintained for weeks, and possibly for months, the severity of the injury and the primary symptoms forming the surgeon's best guide to a decision. This careful line of treatment would also be adopted under the wholesome dread of exciting, by sins either of omission or of commission, the one common complication which the surgeon should ever have before him—namely, an inflammatory action in the injured organ. For

experience speaks in no feeble terms that this action
is readily started and with difficulty quelled, and
that it is by such inflammatory changes in the
injured brain that most head cases, simple or severe,
are brought to a fatal termination. Mr. Hilton
recognised this necessity nearly thirty years ago,
for he taught " that recognised lesions of the brain
and its membranes, associated with blows upon the
head (whether the cranium be fractured or not), do
not generally, or as a principle of treatment, obtain
that extent of mechanical rest which is consistent
with the expectation of perfect and complete structural repair. This error in the treatment of such
cases is one of the chief sources of the diseases of
the brain and its membranes which are met with in
practice. . . . In cases of injury of what may
be called the coarser structures, with more simple
functions attached to them, we see that without
perfect restoration of the structures their functions
are not efficiently performed, and if used too early
and too much they become painful and assume a
chronic inflammatory condition. Such soft parts
require weeks or months for their repair. Surely,
then, we ought not to deny the necessary and pro-
portionately much longer time for the repair of the
more delicate brain tissues; a repair, be it remembered, which cannot be accomplished by any direct
aid from the surgeon, but only by Nature herself
employing her chief agent—Rest." It is a pleasure
to me to be able to quote these apt sentences, framed

by a former teacher and colleague upon this important subject—although many years have passed since they were uttered,—to support the views it has been my privilege to bring before you—views, I may say, that I have for long taught at Guy's, and have reason to believe with some advantage.

Should symptoms of intracranial irritation or inflammation show themselves, they should be dealt with actively, as, from the nature of the brain and its membranous coverings, the process once started soon spreads. In the early stage the application of cold to the head by means of a Leiter's metallic tube is the most efficient local, and free purgation the most effective general, means, with very low diet. If the inflammatory action is great, a free bleeding from the jugular vein or from the arm is strongly to be advocated, and this operation may in many cases be repeated with much advantage. I am convinced I have saved some lives by this treatment. In chronic cases the value of mercury taken internally cannot be doubted.

These, then, are the common lines upon which the treatment of the common factor of all cranial injuries, simple or severe cerebral injury, should always be based; and they should likewise form the lines of treatment of all its complications. Thus, if a simple fracture of the vertex, base, or of both, complicates the case, the treatment is the same. The cerebral injury needs the surgeon's care, and not the fracture, which will take care of itself. A cranial fracture will heal in the same way as other fractures, but it will take a much longer period; and fractures of the base of the skull are apparently amongst the slowest. In specimens 1084[52] (eighty-four days), 1084[55] (ninety-one days), and 1084[56] (eight years), of the Guy's Museum, and in others of our own College museum, this point is indicated. The fractures of the skull will, at any rate, probably heal sooner than the cerebral injury will be repaired. The treatment for the latter will consequently have to be continued after the fracture has healed. If the fracture be but *slightly depressed*, whether simple or compound, and it appears only as a *fissure*, the case had probably better primarily be left alone, and dealt with secondarily on the smallest indication of cerebral trouble; for in these cases there is rarely comminution of the inner table, and consequently nothing in the form of bony spicules to fret and irritate the dura mater, and thus help forward a meningitis. To trephine in order to elevate this form of fissured fracture would therefore be to add another danger to a case in which the form of cerebral injury common to all already exists. If the fracture be *depressed, starred, or comminuted*, whether simple or compound, the elevation of the depressed bone should be the rule of practice, and the removal of all the splintered fragments of the inner table carefully carried out, the object of the operation being more to take away what, if left, must irritate the dura mater, and so add to the existing harm, than to relieve the depression. This operation should be performed as much in simple as in compound fractures, for the condition of the bones is the same in both ; and with our modern treatment of wounds the danger of the operation in simple fracture is not materially increased. In *punctured fractures* the operation of trephining, undertaken with the object of removing the broken and displaced fragments of bone, should be a rule of practice never to be deviated from. The depressed and comminuted inner plates of bone to a certainty, if left, at a late if not early period of the case, irritate the brain and its coverings, and so set up an encephalitis.

How far the presence or absence of what are called brain symptoms should influence a surgeon in his decision as to surgical interference in the different forms of depressed fractures of the skull we have been considering has been much argued. For the surgeon is quite unable in bad cases of cranial injury to differentiate symptoms, and to say how far those that are present in any individual case are due to the common factor—cerebral injury—which resulted from the force that produced the fracture, or how far they are caused by the depressed and fractured bone. But I am not sure that this is a point of much practical importance, for in a bad case of fracture of the skull cerebral injury is probably already severe, and the operation of elevating depressed bone and of removing comminuted fragments is not likely to aggravate the trouble; whereas in a less severe example in which the cerebral injury is likely to be less serious, the existence of depressed bone and of comminuted fragments must act injuriously, and should consequently be removed. Under all circumstances, it is consequently the surgeon's duty to remove whatever sources of trouble the presence of a depressed fracture may bring to an already serious case of cerebral injury.

The operation of trephining or of elevation of bone in depressed fracture is called for more with the object of removing from the brain what may or will be sources of local irritation rather than with any view of removing the effects of the depressed bone; for it is well recognised that, *per se*, a large area, and a considerable amount, of depressed bone are required to bring about symptoms of compression in an otherwise uninjured brain. Again, it is well known that a considerable extravasation of blood upon the surface of the brain, probably five or six ounces, whether between the bone and dura mater or in the cavity of the arachnoid, is required to bring about marked evidence of its presence, in the form of paralysis from compression. The rupture of the middle meningeal artery is a special complication of cranial injuries, or of fracture of the skull, but I do not propose to discuss it here at any length. I cannot, however, pass it by without referring to the exceedingly able article upon the subject pub-

lished in the 'Guy's Hospital Reports' for 1886,
vol. xliii, p. 147, by my friend and colleague, Mr.
H. A. Jacobson, since it contains not only a masterly
account of every case on record up to date, but also
a summary of the whole subject, which claims the
close attention of every surgeon.   His summary is
as follows :

1. That the violence which causes middle meningeal hæmorrhage is often slight, and that in these
cases no fracture may be present.

2. That where there is a fracture, it is often a
mere fissure, and may involve the internal table only.

3. That the history of the case, and, above all,
an interval of lucidity or consciousness, are invaluable,
the latter being worth all the other symptoms put
together.

4. That the symptoms of compression are in some
cases deferred; that their onset may be then very
sudden and rapidly fatal, failure of breathing being
a marked feature.

5. That in those cases where the history is deficient, especially as to any interval of lucidity, and
where it is difficult to be certain about the existence
of hemiplegia, dilatation of the pupil on one side,
that side corresponding to the clot, is a sign of
great value.   The explanation of this sign, that the
third nerve is being pressed upon by a clot large
enough to reach into the middle fossa, we owe to
Mr. Hutchinson, with whose name in future this
condition of the pupil should be associated.

6. That after trephining, exposure and partial removal of the clot, very severe hæmorrhage may set in and prove difficult to arrest.

7. That in severer cases laceration or contusion of the brain are only too frequently complications.

This latter conclusion consequently links this special class of cases with the more general number of cranial injuries, and enables the surgeon to look upon them as a whole in the light in which I have now placed them before you.

There are many other questions in the surgery of cranial injuries which require elucidation, but time will not allow me to bring them under your notice. The points and questions I have selected are such as I believe to be most important to enable the younger surgeons and practitioners to read rightly the manifold and somewhat puzzling phenomena which severe cranial or cerebral injuries exhibit, and I have some confidence in the belief that, if the views I have expounded were accepted, the teaching and understanding of cranial or cerebral injuries would be greatly simplified. In conclusion, I must ask you who have listened to me so patiently and kindly to think over the questions I have ventilated, and to accept from me my warmest acknowledgments of the honour you have conferred upon me by allowing me, as your Hunterian Professor of Surgery, to deliver these lectures from this chair.

INDEX.

PAGE

Abscess and tension . . . 32
— in bone, cleansing of . . 73
Appearance of tense tissues . . 6

Bleeding, local, in tension . . 37
Bone, circumscribed abscess in . 83
— — Brodie's cases of . . 83
— — Ferguson on . . . 84
— tumours of 10
— — tension in 11
— inflammation of, and tension . 43
— — in scalp wounds . . 123
— vascular supply of . . . 44
— chronic inflammation of . . 65
— treatment of . . . 65–88
Brain changes in head injuries . 100
— bruising in head injuries . 104
— laceration in head injuries . 104
— lasting nature of . . . 120
Bruising of brain 104

Cerebral injury, cases of . . 109
Concussion, brain changes in . 102
— — in old cases . . . 108
— meaning of word . . . 100
— its vagueness 121
— how produced . . . 101
— complications of . . . 102
— cases illustrating . . . 109
Conclusions on tension . . . 36
— — in inflamed bone . . 95
Cleansing abscess cavity in bone . 73
Cranial bones differ in thickness . 99
— — influence of this fact in
injuries 99
Cupping in tension . . . 37

Death of tissue from tension . 23
Diagnosis of tension . . . 5

Diagnostic value of local pain . 50
Drilling inflamed bone . . . 51
— — in joint complication . . 80
— — in chronic cases . . 74, 92

Effects of tension on inflamed
tissues 22
— of inflammation on repair . 34
— concussion of the brain . . 108

Fractures of skull . . . 107
— — of base 109
— — treatment of . . . 140

Glaucoma and tension . . . 29

Hæmorrhage, cause of tension . 15
— — in testicle 17
— — in eyeball . . . 16
— — in bulb of urethra . . 19
Head injuries 98
— how to be estimated . . 99
— treatment of 123
Heat of part sign of local inflammation 49
— how to test it . . . 49
Hip disease, drilling bone for . 80

Incision down to bone, value of . 50
Inflammation and tension . . 21
— — in bone 45
— effects of acute . . . 45
— — of chronic . . . 46
— of bone in head injuries . . 123
Intracranial inflammation, treatment of 139

Leeching in tension . . 37

Necrosis 45

10

146 INDEX.

PAGE

New growths, tension from . . 7
Nocturnal pains and tension 22–49

Ostitis in head injuries . . 123
— illustrative cases . . . 126

Pain, value of local, as sign of
  tension 36
— in inflamed bone . . . 48
— caused by tension . . . 14
— from hæmorrhage . . . 15
— — tumours 21
— — inflammation . . . 21
— — suppuration . . . 32
Palpation in tension . . . 5
Partial relief of tension, effects of 27–8
Periostitis and ostitis, cases of . 53
Pott, observation of, on head
  injuries 124
Puncture in tension . . . 39

Relief of tension as a principle of
  practice 26
Repair and inflammation incompatible 33
Rest in head cases . . . 137

Scalp wounds in head injuries,
  treatment of . . 123, 132
Suppuration and tension . . 31
Symptoms of tension . . . 5

Tapping joint to relieve tension . 27
Tension 1
— its meaning 2
— its causes 2
— its effects . . . 3, 4
— general conclusions on . . 36

PAGE

Tension, influence of tissue on . 3
— when acting rapidly . . 3
— — slowly . . 3
— from blood 4
— — new growths . . . 7
— — inflammation . . 4, 21
— symptoms of 5
— diagnosis of 5
— death of tissue from . . 23
— treatment of . . . 26, 37
— in glaucoma 29
— in inflamed tooth pulp . . 29
— in ear affection . . . 30
— in suppuration . . . 31
— in wounds 32
— in peritonitis 39
— in ostitis 43
— in periostitis 43
Treatment of acute inflammation
  of bone 50
— of chronic inflammation of
  bone 65
— of ostitis in head injuries . 132
— of tension 37
Trephining inflamed bone . . 51
— in chronic cases . . 64, 88
— in abscess . . . 69, 83
— in cranial ostitis . . . 133
— in head injuries . . . 140
— in meningeal hæmorrhage . 143
Tumour beneath temporal fascia . 9
— in median nerve . . . 9
— in brachial plexus . . . 10
— in shaft of tibia . . . 11
— in lower jaw 13

Vascular supply of bone . . 44

Wounds and tension . . . 32